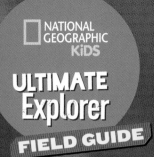

British Insects

NATIONAL GEOGRAPHIC

WASHINGTON D.C.

Contents

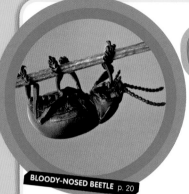

BLOODY-NOSED BEETLE p. 20

CRANEFLY p. 39

FIREBUG p. 55

RUBY-TAILED WASP p. 80

PURPLE EMPEROR p. 89

WHITE-FACED DARTER p. 109

INSECTS ARE INVERTEBRATES (animals without backbones) that have been around for hundreds of millions of years. They are the largest and most diverse group of animals found on Earth today.

Arthropods

Insects belong to a group of animals called arthropods. Arthropods don't have internal skeletons and backbones like people do. Instead, they have an exoskeleton. An exoskeleton is a hard covering on the outside of the animal's body.

Some exoskeletons are thick and hard; others are thin and light. Some let water and gases in; others keep them out. Adaptations in the exoskeleton are one reason that insects have become the most successful animal group on Earth.

Arthropods have two other notable features that make them easy to identify. Their bodies are segmented, or divided into parts. And their arms and legs are jointed. The word *arthropod* even means "jointed leg."

What is an insect?

Insects are just one type of arthropod. Others include spiders, crabs, lobsters, centipedes and scorpions. What makes an insect different?

- Insects have six legs. Other arthropods have more.

- An insect's body is divided into three main parts: the head which contains the mouthparts, antennae and eyes; the thorax where legs and wings are attached; and the abdomen at the end of the insect's body containing its reproductive, digestive and respiratory systems. Other arthropods have one or two sections.

- Insects usually have wings. No other arthropods have wings.

Bugs v insects

You often hear people referring to insects as "bugs"—we've even done it on the front of our book!—but this isn't always correct. All true bugs are insects and they have six legs and three segmented body parts, but in the same way that not every arthropod is an insect, not every insect is a bug! Bugs are a large family of insects with the scientific name Hemiptera.

Metamorphosis

All insects undergo a change in form during their lifespan. This process is called metamorphosis. While all insects go through metamorphosis, the way it happens differs considerably between insect groups (see pages 8–9).

How insects use their body parts

Just like people, insects have the senses of sight, hearing, taste, touch and smell, but their senses work very differently than ours do.

- **Sight:** Insects can have simple eyes, compound eyes, both or neither. Simple eyes distinguish between dark and light. Compound eyes can help insects see a wider range of colors or see in all directions.

- **Hearing:** Many insects can't hear. Those that can hear use different body parts to do so. Crickets hear with a structure called a tympanum on their front legs. Some moths can also hear.

- **Taste:** Most insects taste with their mouthparts or their feet. Bees have taste receptors on their antennae.

- **Touch:** Insects have small hairs on their bodies. There's a nerve at the base of each hair that can feel objects and movement in the air around them.

- **Smell:** Insects have receptors on their antennae that detect and gather molecules of odours in the air. That's how they smell.

GREAT GREEN BUSH-CRICKET p. 110

Where they live

Insects have adapted to live in all types of environments. They live inside and outside, in hot places and cold places. Some insects even live on you!

- **On plants:** Caterpillars are the larvae of butterflies and moths. They are often found on the underside of leaves. Aphids are small insects that live on stems and leaves. If you happen to spot aphids, you will probably see ladybirds, too—aphids are one of their favourite foods!

- **In the air:** If you're outside, don't forget to look up. You can see all kinds of insects flying around. At night, many different insects are attracted to lights. Fireflies and a few other species glow in the dark.

- **On the ground:** Many insects live under rocks and logs. Some, such as Red Soldier Beetles or Green Tiger Beetles, are common in gardens and flower beds hunting for smaller insects like aphids that are their prey. The larvae of these beetles live at the base of the plants near the ground where they eat snails and slugs.

- **In water:** Pond Skaters live on top of ponds. They have long legs covered with tiny hairs which allow them to gently balance on the water's surface.

- **In the home:** It's just as easy to find insects in and around your home. Wasps often build their nests under the eaves of houses. Even when they are not nesting close by, bees and wasps will fly inside when the windows are open in the summer. Elsewhere in the home, you can often find a number of other insect species. Good places to look are near food, particularly fresh or decaying fruit, or in the corners of dark cupboards.

Metamorphosis

THROUGHOUT THE ANIMAL KINGDOM many species take on quite different appearances as they grow, but there are few creatures that undergo such fantastic changes as insects. This process of change is called metamorphosis and while it is not unique to insects, the resulting differences between juvenile and adult insect species can be the most dramatic.

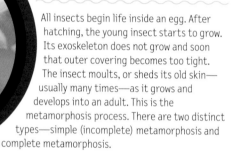
7-SPOT LADYBIRD (larva) p. 17

All insects begin life inside an egg. After hatching, the young insect starts to grow. Its exoskeleton does not grow and soon that outer covering becomes too tight. The insect moults, or sheds its old skin—usually many times—as it grows and develops into an adult. This is the metamorphosis process. There are two distinct types—simple (incomplete) metamorphosis and complete metamorphosis.

Simple (incomplete): Species like grasshoppers and dragonflies undergo simple metamorphosis. Simple metamorphosis has three stages: egg, nymph and adult. The nymph usually resembles a small adult in shape and the insect remains active throughout its growth period.

Complete: Species like butterflies, moths and beetles go through complete metamorphosis. Complete metamorphosis has four stages: egg, larva, pupa and adult. The larva usually bears no resemblance to the adult insect. During the larva stage the insect will hungrily eat as much food as it can before entering an inactive pupa phase. It exits this pupa phase as a fully developed adult.

7-SPOT LADYBIRD (adult) p. 17

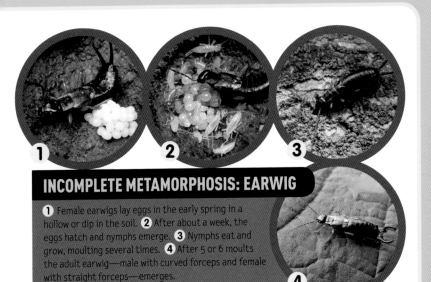

INCOMPLETE METAMORPHOSIS: EARWIG

1 Female earwigs lay eggs in the early spring in a hollow or dip in the soil. **2** After about a week, the eggs hatch and nymphs emerge. **3** Nymphs eat and grow, moulting several times. **4** After 5 or 6 moults the adult earwig—male with curved forceps and female with straight forceps—emerges.

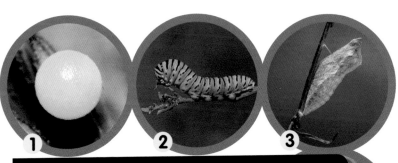

COMPLETE METAMORPHOSIS: SWALLOWTAIL BUTTERFLY

1 A caterpillar develops inside an egg. **2** After 3 to 15 days, the caterpillar hatches from the egg. The larva feeds and grows for about two weeks. It moults 5 times. **3** Then the caterpillar turns into a pupa. **4** 2 weeks later, an adult butterfly emerges.

SEARCHING FOR
Insects

WE'VE ALREADY DESCRIBED how insects are found all over the place, but to find the widest range of insects, especially those rare or shy ones, you'll have to leave your house and go searching.

RED ADMIRAL p. 90

When to look for insects

It's pretty easy to find insects during the summer months. Some can also be seen in the spring and autumn. It's much harder to spot them outside during the cold winter months. There are several reasons for that. Firstly, some insects pass the winter months in the larva, pupa or nymph stage. The insects are there, but they're not in the adult form that people are most likely to recognise. They may also be buried underground. Insects that are adults at this time of year have different ways of dealing with winter. Many butterflies—including Red Admirals and Painted Ladys— migrate to warmer climates to avoid the cold temperatures that would surely kill them. Other insects—like ladybirds—hibernate through the winter.

Play it safe

If a wasp, bee, or hornet stings you, it will hurt. If a mosquito or flea bites you, it will itch. Usually, with the insects found in the UK, the reaction is mild, but if you're allergic to the bite or sting, you may need medical attention. If you know you have an allergy when you go out looking for insects, carry an emergency epinephrine kit with you. Some insects carry disease. In some parts of the world, a mosquito bite can give you malaria or some other infection. Most UK insect bites or stings will not lead to health problems, but if you do feel funny or notice a reaction after a bite or sting always tell a grown-up.

Common Wasp p. 78

The best way to stay safe when observing insects is to be prepared. Wear long trousers, long sleeves and sturdy shoes. Try to dress in earth colours like tans and greens to help you blend into the surroundings— this will help you get closer to insects you want to see and won't attract those that you'd prefer to avoid! Use insect repellent and avoid wearing hairspray, perfume, or other scented products. Wear a hat on a sunny day. Most importantly, try not to bother individual insects, and NEVER disrupt a hive or nest.

Protecting insects

Insects are highly adaptable, and they make up about 80 per cent of the animal diversity on Earth. Because of this, many people don't think about protecting and conserving insects when they clear land and use natural resources. They should.

Insects are tiny, but they play a major role in the food chain. Insects provide food for all kinds of animals, including other insects. They pollinate crops, make honey, make silk, and clean up the environment. In some parts of the world, people even eat insects for food. If insects were to disappear, the impact would be enormous.

When you are out and about looking for insects be careful not to disturb what may be a very fragile home to them. Remember the old saying that: "they are more scared of you, than you are of them". Unless you are looking at the bravest ant, it is probably true. Tread carefully. Try to avoid stepping on any other little creature—insect or other!

If you enjoy looking at insects, you might want to actively help to conserve and protect their habitats. You might even want to create some insect habitats yourself. For example, some people build butterfly gardens. To do this, find a sunny location, make sure it's sheltered from the wind, then choose plants that attract butterflies and others that caterpillars can eat.

You can also help insects by reminding adults not to use chemicals on lawns or near bodies of water where insects might live. Some insects are pests, and people use pesticides to keep down their populations. Unfortunately, other insects can be affected, too. Encourage your friends and family not to use chemicals unless they absolutely have to.

✓ CHECKLIST FOR FINDING INSECTS IN THE FIELD

The best way to learn about ladybirds, shield-bugs and pond skaters, is to go out and take a look. Here are some things to bring on your search for insects:

✓ BINOCULARS OR MAGNIFYING GLASS Insects come in many sizes, but even the biggest have some parts that are hard to see with the naked eye. A magnifier will help you get a good look if you can get close enough. If it's flying in the distance, binoculars are better.

✓ GET A GOOD GUIDE Take this book with you on hikes, bike rides and picnics. See how many insect species you can identify. As your interest builds you might want to get a more focused guide that includes all the species in your area.

✓ MAKE A NOTE OF IT
Pack a small notebook and pen or pencil in your backpack. You'll want to keep a record of the species you see and when and where you find them. You can even make quick sketches that will help you identify them later.

✓ PROPER CLOTHING Many insects—including some that you'd prefer to avoid—are attracted to bright colours. Wearing earth colours like tans and greens helps you blend into your surroundings. Long trousers, long sleeves and sturdy shoes will protect against nettles, thorns and ticks. Wear a hat for sun protection.

Turn the page to get started finding and identifying insects!

HOW TO USE This Book

GO SEARCHING! You may find insects in your garden or in the local park, or you may have to go out on a nature ramble to find more exotic habitats—like ponds and marshes. Wherever you do find insects, check this guide to learn about them!

WHITE ADMIRAL p. 94

Insect Entry

THIS IS WHERE YOU'LL FIND THE INSECT'S COMMON NAME

HERE IS THE INSECT'S SCIENTIFIC NAME, ITS AVERAGE SIZE AND DETAILS ABOUT ITS HABITAT AND RANGE

BUFF-TAILED BUMBLEBEE

Bombus terrestris
LENGTH 20–22 mm · HABITAT Gardens and parks · RANGE Common in lowland areas throughout the UK and Ireland from March to August
COMPLETE METAMORPHOSIS

OUR LARGEST BUMBLEBEE, the Buff-tailed Bumblebee is named after the queen bee's buff-coloured tail. The worker bees have white tails with a faint buff line separating them from the rest of the abdomen. The Buff-tailed Bumblebee visits many different types of flowers for pollen and nectar; it has a short tongue, however, so prefers open, daisy-like flowers. Look for the orange or golden collar and second abdominal segment. This bumblebee nests underground in large colonies of up to 600 bees, often using the old nests of small mammals.

THIS TEXT GIVES GENERAL INFORMATION ABOUT THE SPECIES, INCLUDING INSECT BEHAVIOUR AND SOME SURPRISING FACTS

BUFF "TAIL"

INSECT Inspector!

Buff-tailed Bumblebees are known as "nectar robbers"; if they come across a flower that is too deep for their tongue, they bite a hole at its base and suck out the nectar.

DISCOVER FUN FACTS OR IDENTIFYING TIPS FROM EXPERTS

COMMON CARDER BEE

Bombus pascuorum
LENGTH 13 mm · HABITAT Gardens, farmland, woodland, hedgerows and heaths · RANGE Widespread, from March to September
COMPLETE METAMORPHOSIS

THE COMMON CARDER BEE is a fluffy, brown-and-orange bumblebee, sometimes displaying darker bands on the abdomen. It is one of our most common bumblebees and nests in cavities, such as old mouse runs and birds' nests. It is a social insect; nests may contain up to 200 workers. Common Carder Bees are one of several "long-tongued bees" that feed on flowers with long tubular florets, such as Heather, Clover and Lavender.

ORANGE FURRED THORAX

5 OR MORE ORANGE STRIPES ON THE ABDOMEN

IDENTIFY AN INSECT BY LOOKING FOR THESE BASIC FEATURES. CAN YOU NAME THE CORRECT SPECIES IN 10 SECONDS?

FAMILY NAME

66 HYMENOPTERA

Special features called Insect Reports give you a closer look at unique insect habitats, strange behaviours or interesting features.

A CAPTION DESCRIBES THE MAIN PHOTO

LEARN ABOUT THE DIFFERENT SPECIES THAT CAN BE FOUND THERE

A TEXT BLOCK GIVES GENERAL INFORMATION ABOUT THE THEME OF THESE REPORT PAGES

Classification

INSECTS ARE ALL AROUND YOU. More than one million different species have been identified and many more have yet to be discovered. According to some estimates, there are nearly 10 quintillion (10,000,000,000,000,000,000) insects in the world.

That's a lot of insects to keep track of. To help people understand insects, scientists divide them into groups based on their common traits. This process is called scientific classification. All living things are categorised in this way.

Scientific classes are usually based on the characteristics or behaviours of a specific insect group. Latin terms for the characteristic or behaviour are usually used for the scientific name. For example, the Latin word Palaeoptera means "old wing" and is used to describe a group of insects that have large membranous wings which they can't fold—mayflies, dragonflies and damselflies belong to this group. The insects in this book have been grouped into their classes to help you identify them.

WORMWOOD MOONSHINER p. 35

INSECT REPORT: BEETLE MANIA

Burying Beetle

Nicrophorus vespillo

Known as the undertakers of the animal world they are found near small animal corpses—they can smell them from MILES away! They often move the corpse and the female will use the burial spot to lay her eggs (see page 21).

Beetles live everywhere on Earth except for the oceans and Antarctica. By number of species, they are the largest group in the animal kingdom—nearly a quarter of all animal types on Earth are beetles! Like all insects, beetles have three main body parts: head, thorax, and abdomen. Beetles' legs and wings are attached to the thorax. Their front wings are thick and hard, covering and protecting most of a beetle's body, while the membranous hind wings are used to fly.

 SOME BEETLES (LIKE THE STREAKED BOMBARDIER BEETLE) CAN SECRETE CHEMICALS THAT CAUSE PAIN, RASHES OR ITCHING, BUT MOST DO NOT.

Green Tiger Beetle

Cicindela campestris
This sun-loving beetle is most likely to be seen scuttling rapidly in search of insects, flying with a buzzing sound when disturbed. This predator is well-equipped to tackle its prey, with a ferocious set of jaws and long legs that make it one of the fastest insects (see page 25).

Common Bombardier Beetle

Brachinus crepitans
The "crepitans" part of this ground beetle's scientific name means "crackle" in Latin and comes from the noise it makes. This beetle does not have the explosive defence mechanism of its relative the Streaked Bombardier Beetle (see page 31).

BEETLES UNDERGO COMPLETE METAMORPHOSIS. A LADYBIRD FOR EXAMPLE, GOES THROUGH FOUR PHASES: THE EGG; THE LARVAL STAGE, DURING WHICH THE LARVA UNDERGOES A SERIES OF MOULTS; THE PUPA, IN WHICH THE LARVA DEVELOPS INTO AN ADULT; AND THE ADULT PHASE, DURING WHICH THE FEMALE LAYS EGGS IN BATCHES OF UP TO 40—STARTING THE WHOLE PROCESS AGAIN.

COMMON COCKROACH

Blatta orientalis LENGTH **15–30 mm**
● HABITAT **Heated buildings, including greenhouses, and in coal mines and sewers** ● RANGE **Very common, from January to December**
INCOMPLETE METAMORPHOSIS

THE FOREWINGS OF the male Common Cockroach cover most of the abdomen, but those of the female are reduced to tiny flaps just behind the pronotum. Neither sex can fly, but they are fast and agile runners. The Common Cockroach—also known as the 'Black Beetle' and the 'Oriental Cockroach'—originally came from North Africa or Asia. It lives as a scavenger in buildings and on rubbish dumps.

LONG, SPINY LEGS

10s. spotters

INSECT inspector!

You will often hear it said that cockroaches are fantastic survivors and it is true that they can tolerate very high and very low temperatures, but did you know they can even survive—at least for a few days—without a head? This is because they breathe through little holes in their bodies and being cold-blooded, they can last several weeks without food. In fact—if it was not eaten by a predator—a headless cockroach would probably survive for about a week, before it would eventually die of thirst.

HARD, PROTECTIVE WINGS

2-SPOT LADYBIRD

Adalia bipunctata **LENGTH 4–6 mm** ▪ **HABITAT Parks,
towns and gardens** ▪ **RANGE Very common throughout the UK
though less so in the far north, from March to October**
COMPLETE METAMORPHOSIS

THE 2-SPOT LADYBIRD is a
medium-sized ladybird, usually red
with two black spots on the wing
cases, but it also comes in a variety of
other colour forms, right through to
black with two red spots. The adults
hibernate over winter in bark, or sometimes in
houses, congregating in large numbers. Adults and
larvae feed on aphids, making them a friend in the garden.
Adult ladybirds can survive for up to a year.

BLACK LEGS

WHITE SPOTS
ON FACE

BLACK HEAD
(NOT MOTTLED)

7 SPOTS

7-SPOT LADYBIRD

Coccinella septempunctata **LENGTH 6–8 mm** ▪ **HABITAT Open fields,
grassland, marshes, agricultural fields, gardens and parks** ▪ **RANGE Widespread
throughout the United Kingdom usually seen March to October**
COMPLETE METAMORPHOSIS

THE 7-SPOT LADYBIRD is easily recognised by
its red wing cases, dotted with a pattern of
seven black spots; it also has a familiar black-
and-white-patterned thorax. The 7-Spot
Ladybird is "the" ladybird that everyone is familiar
with. A virtually ubiquitous inhabitant of gardens
and parks, it turns up anywhere there are aphids to
feed on. Adults hibernate in hollow plant stems and cavities,
sometimes clustering together in large numbers. The 7-Spot Ladybird is also
a migratory species: large numbers fly in from the Continent every spring,
boosting our native population. The life cycle of a ladybird consists of four
phases: the egg; the larval stage, during which the larva undergoes a series
of moults; the pupa, in which the larva develops into an adult; and the adult
phase, during which the female lays eggs in batches of up to 40.

HARLEQUIN LADYBIRD

Harmonia axyridis
LENGTH 8 mm • HABITAT Towns and gardens
• RANGE Widespread in England and Wales and
spreading into Scotland, from April to October
COMPLETE METAMORPHOSIS

THE HARLEQUIN LADYBIRD
is one of the most invasive
insect species in the world. It is
a non-native species, originally
from Asia, and is a voracious
predator—it can out-compete our
native species for aphid-prey and will
also eat other ladybirds' eggs and larvae.
Over 100 different colour patterns have been
recorded which makes the Harlequin Ladybird difficult to identify,
especially from the 7-Spot Ladybird, which is also variable. Its head
has an obvious white triangle in the centre.

WHITE
TRIANGLE

INSECT Inspector!

In North America, the Harlequin Ladybird is sometimes known as the
"Halloween Bug" because it gathers together in enormous numbers during
the late autumn, sometimes invading people's homes.

EYED LADYBIRD

Anatis ocellata LENGTH 8–10 mm • HABITAT Gardens and
parks • RANGE Common, March to October
COMPLETE METAMORPHOSIS

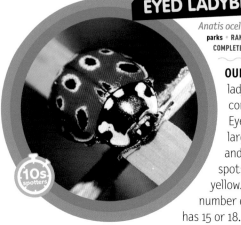

OUR LARGEST LADYBIRD, the Eyed
ladybird is usually found on, or near,
conifers—especially pine trees. The
Eyed Ladybird is unmistakeable: it is
larger than all the other ladybirds
and is the only one that has 'eyed'
spots—black spots ringed with
yellow. Its wing cases are red. The
number of spots can vary but it usually
has 15 or 18.

ASPARAGUS BEETLE

RED THORAX

Crioceris asparagi LENGTH 5–8 mm • HABITAT Rough ground and gardens—wherever wild or cultivated asparagus grows • RANGE Widespread throughout the UK, from January to December
COMPLETE METAMORPHOSIS

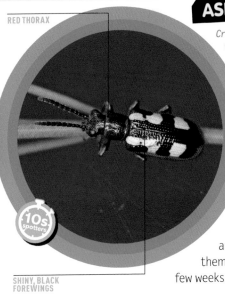

THE COMMON ASPARAGUS

Beetle is metallic blue-black in colour with cream or yellow spots on its red-bordered wing cases. The larvae are black-spotted and grey. The adult beetles and the larvae strip the needle-like leaves off asparagus plants, chew the spears and lay generous amounts of eggs on them. The larvae feed on the plants for a few weeks, then drop to the ground to pupate.

SHINY, BLACK
FOREWINGS

BLACK-OIL BEETLE

Meloe proscarabaeus
LENGTH Up to 30 mm • HABITAT Meadows and fields
• RANGE Widespread and locally common throughout the UK, from February to June
COMPLETE METAMORPHOSIS

THE BLACK-OIL BEETLE is flightless, bulky, has a bulbous abdomen, a large flat head and a metallic sheen. It is named for the smelly, oily fluid it produces when threatened. The male is smaller than the female and has conspicuously kinked antennae. Adults are plant-eaters, but the grubs live in the nests of solitary bees. They eat the bee eggs as well as the stored pollen and nectar.

FLAT, BLACK
HEAD

BLOODY-NOSED BEETLE

Timarcha tenebricosa
LENGTH 20–23 mm • HABITAT Grassland, heathland and along hedgerows • RANGE Found in South and Central England, and Wales from April to September COMPLETE METAMORPHOSIS

THIS BEETLE IS generally the largest of our UK leaf beetles and gets its name because, when threatened, it oozes a distasteful red liquid from its mouth that wards off would-be predators. It is a large, domed, black beetle with a bluish sheen and long legs, that can often be seen moving slowly across paths or through grass. The line running down its back gives the impression of separate wing cases, but they are in fact fused together and this beetle is flightless. Adults are mostly active at night.

THICK ANTENNAE

DOMED SURFACE

BLUE GROUND BEETLE

Carabus intricatus LENGTH 24–35 mm • HABITAT Ancient woodland • RANGE A very local and rare species. Established in Dartmoor, Bodmin Moor and a coastal site in South Wales COMPLETE METAMORPHOSIS

THE BLUE GROUND BEETLE is a large, distinctive beetle with stunning metallic blue markings, long legs and sculpted wing cases. It was thought to be extinct in the UK until it was rediscovered in Dartmoor National Park in 1994. Both larvae and adults are mainly nocturnal, and are thought to feed largely on tree slugs. Adults are found under bark on dead wood, and under rocks and they like damp, rotten, moss-covered wood. It is thought that it may take two years for the Blue Ground Beetle to complete its life cycle.

SHINY BLUE-BLACK MARKINGS

INSECT Inspector!

Blue Ground Beetles have been found to have a taste for dog food.

COMMON BOMBARDIER BEETLE

Brachinus crepitans • LENGTH 5–10 mm • HABITAT Dry grassland, especially on limestone, where it usually hides under stones • RANGE Found in southern England and southern Wales, particularly in coastal areas, from May to June
COMPLETE METAMORPHOSIS

ORANGE-RED HEAD

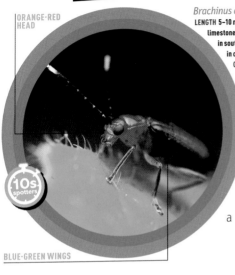

BLUE-GREEN WINGS

THIS GROUND BEETLE is one of two Bombardier Beetles found in the UK—both types are rare, but this one is more abundant than the Streaked Bombardier Beetle (see page 31). The head and thorax are orange-red while its wing cases are a metallic blue-green colour.

Nicrophorus vespillo LENGTH 25–35 mm • HABITAT Coastal, farmland, grassland, heathland and woodland habitats, towns and gardens • RANGE Widespread throughout the UK including Northern Ireland, from April to October
COMPLETE METAMORPHOSIS

BURYING BEETLE

BOBBLE AT ANTENNAE TIP

BURYING BEETLES (also called Sexton Beetles) are known as the undertakers of the animal world and can be found wherever there are small animal corpses—they can smell one from a mile away! Males and females pair up and move the corpse to a suitable spot for burial. The female will then lay her eggs on the soil immediately above the body. Once the eggs have hatched, the body is used as food for the larvae. Burying Beetles can be identified by the brightly coloured bands of orange-red on their wing cases and bright orange bobbles on the ends of their antennae.

CLIFF TIGER BEETLE

Cylindera germanica
LENGTH **Up to 45 mm** ◦ HABITAT **Soft stone and sandy cliffs**
◦ RANGE **South-facing cliffs of Dorset, Devon and the Isle of Wight**
COMPLETE METAMORPHOSIS

THE CLIFF TIGER BEETLE is the cheetah of the insect world: with its long legs, it speeds after prey before catching it in its large and fearsome jaws. Its larvae live in burrows in damp sand, where they lie in ambush for unsuspecting victims which are grabbed and pulled in to be eaten. Cliff Tiger Beetles are solar-powered, which is why they are only found on south-facing cliffs in the UK. Many invertebrates that are restricted to soft cliffs are solar-powered (the technical term for this is "thermophillic", or "warmth-loving").

FLAT APPEARANCE

COBWEB BEETLE

BRISTLY BODIED LARVAE

Ctesias serra
LENGTH **10 mm** ◦ HABITAT **Ancient gnarled broadleaf trees** ◦ RANGE **Throughout England but rare in Wales and Scotland, from January to December**
COMPLETE METAMORPHOSIS

THE CATERPILLAR-LIKE larvae of the Cobweb Beetle steal the remains of dead bugs from spiders' webs. They live on ancient gnarled broadleaf trees, where they can be found all year round under loose webby bark and around the fringes of spiders' webs—close to their dinner! The long bristles covering their bodies protect the larvae from attacks by spiders. When threatened, the larvae vibrates its long bristles—spiders just can't get their fangs through all those hairs! The Cobweb Beetle is threatened due to its rare and declining habitat.

COMMON COCKCHAFER

LEAVED ANTENNAE

Melolontha melolontha **LENGTH 20–30 mm**
- **HABITAT** Parks and gardens, grassland and woodland
- **RANGE** Widespread, but rarer in the north

COMPLETE METAMORPHOSIS

THE COMMON COCKCHAFER is a large, rusty-brown beetle also known as the "May Bug" because it emerges in large numbers during spring. Males can easily be distinguished from females by counting the number of "leaves" on their remarkable antler-like antennae—males sport seven "leaves" while females have only six. Its larvae live underground for several years, eating the roots of grasses and plants. In May and June, the adults emerge from the soil, often swarming around treetops. Common Cockchafers can be seen at dusk and in the evenings, and are attracted to street lights and lighted windows.

HAIRY THORAX

DEVIL'S COACH HORSE

EXPOSED ABDOMEN

Staphylinus olens **LENGTH 20–30 mm** - **HABITAT** Woods, hedgerows, gardens and many other places: often in sheds and outhouses - **RANGE** Widespread

COMPLETE METAMORPHOSIS

THE DEVIL'S COACH HORSE is a black, medium-sized beetle, with large jaws. It is well-known for curling up its tail in a scorpion-like position when threatened, and emitting a foul-smelling substance from its abdomen. It is an aggressive, carnivorous predator, emerging after dark to prey on slugs and other invertebrates, and using its pincer-like jaws to crush them. Devil's Coach Horse are fast-moving, preferring to run along the ground rather than fly.

SHORT WINGS

INSECT inspector!

Beware—a Devil's Coach Horse can deliver a painful bite, even to a human!

DUNG BEETLE

Anoplotrupes stercorosus
LENGTH 5–30 mm • HABITAT Forest areas, particularly Beech
forests • RANGE Found across the whole country, especially
western Scotland and central Wales
COMPLETE METAMORPHOSIS

DUNG BEETLES CAN bury dung up to
250 times heavier than themselves in
one night. Many Dung Beetles, known as
rollers, roll dung into balls using their
scoop-like head and paddle-shaped
antennae. These are used as a food source
or breeding chambers. Others, known as
tunnellers, bury the dung wherever they find it. A
third group, the dwellers, neither roll nor burrow: they
simply live in manure. Dung Beetles are usually dark in colour,
although some have a metallic lustre.

BROWN
ANTENNAE

RIDGED
WINGS

GLOW-WORM

YELLOW/ORANGE DOTS
ON BODY SEGMENTS

Lampyris noctiluca
LENGTH 15–25 mm • HABITAT Grassy and scrubby places,
including roadside verges and hedgerows: mainly on lime
• RANGE Found in parts of England (particularly the south),
lowland Scotland and Wales, from June to July
COMPLETE METAMORPHOSIS

THE GLOW-WORM IS not
actually a worm, but a beetle.
The male is a light brown,
typical beetle. The larva is
greyish-brown with yellowy-or-
ange triangular markings at the
side of each segment. The females
have no wings and are similar in
appearance to the larva, but are
unmistakable when they light up at
night—emitting a greeny-orange light from their
bottoms! Females climb up plant stems and glow to attract males, who have
large, photosensitive eyes—perfect for scanning vegetation at night.

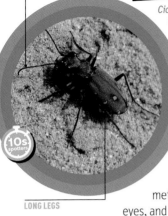

LONG ANTENNAE

LONG LEGS

GREEN TIGER BEETLE

Cicindela campestris
LENGTH 10–15 mm ▪ HABITAT Heaths and other sandy places, including coastal dunes ▪ RANGE Widespread and common, from April to September
COMPLETE METAMORPHOSIS

THIS SUN-LOVING BEETLE is most likely to be seen scuttling rapidly over the ground in search of insects, but it flies with a buzzing sound when disturbed. An agile predator, it is well-equipped to tackle its prey, with a ferocious set of jaws and long legs that give it an impressive turn of speed. The Green Tiger Beetle is a dazzling, metallic-green colour, with purple-bronze legs and eyes, and large, creamy spots on its wing cases.

....

LESSER STAG BEETLE

Dorcus parallelipipedus
LENGTH to 30 mm ▪ HABITAT Woodland, parkland and hedgerows ▪ RANGE Found throughout England and Wales, from May to September
COMPLETE METAMORPHOSIS

FLAT, WIDE HEAD

LARGE JAWS

THE LESSER STAG BEETLE is a large beetle with a broad head and large jaws. The larvae feed on old tree stumps and other rotting wood. Both adults and larvae can be found in the decaying wood of Ash, Common Beech and Apple—where adults mate and lay their eggs. Adults can be seen flying about in the evenings and are attracted to outside lights. The Lesser Stag Beetle is sometimes mistaken for the rarer and larger Stag Beetle.

LILY BEETLE

Lilioceris lilii
LENGTH 6–8 mm ◦ **HABITAT Gardens and anywhere else where plants of the lily family grow** ◦ **RANGE Widespread, most often seen from April to August**
COMPLETE METAMORPHOSIS

THIS BRILLIANT LEAF-EATING beetle is a serious pest of cultivated lilies and also feeds on wild members of the lily family. Adult beetles make rounded holes in the leaves and will also feed on petals and seed pods. Look for the orange grubs, clothed with black slime, feeding on the plants in the summer. There are up to three generations in a year. Adults of the final generation hibernate in the soil.

VIBRANT RED, RIDGED WINGS

LONGHORN BEETLE

Agapanthia villosoviridescens
LENGTH 10–22 mm ◦ **HABITAT Parkland and hedgerows** ◦ **RANGE Widespread across the south and east of England**
COMPLETE METAMORPHOSIS

LONGHORN BEETLES are typically characterised by extremely long antennae, which are often as long as or longer than their body. Most longhorn larvae bore into wood and can cause extensive damage to living trees and, occasionally, to wood in buildings. The female beetle lays her eggs in rotting fungus, living or dead wood. Once the eggs hatch, the larvae begin feeding upon their food source. Once the larvae reach a suitable stage in their development, they begin to pupate and then finally emerge as adult beetles. Adults are often found on flowers or on recently fallen or felled timber.

LONG ANTENNAE

GOLDEN-FLECKED THORAX

PINE WEEVIL

Hylobius abietis
LENGTH **10–13 mm** • HABITAT **Coniferous woods and plantations**
• RANGE **Widespread across England and Scotland**
COMPLETE METAMORPHOSIS

PINE WEEVILS ARE small beetles. The front of the head is drawn out to form a beak with the jaws at its tip. They are dark brown with patches of cream or chestnut hairs arranged in irregular rows on their elytra. Pine Weevils are common in coniferous woods where adults interfere with tree growth by chewing the bark of young shoots. The larvae live in dead tree stumps.

BEAK-LIKE JAWS

POT BEETLE

COLOURFUL
FOREWINGS

SHORT, STOCKY
BODY SHAPE

Cryptocephalus ssp.
LENGTH **5–8 mm** • HABITAT **On trees in forests, particularly Oak and Hawthorn** • RANGE **Common across southern England and Wales**
COMPLETE METAMORPHOSIS

POT BEETLES GET their name from the protective shell-like "pot" that the larvae live in, created using their own droppings! After laying each of her eggs, the female Pot Beetle spends time covering them with a waxy coating and some of her own droppings. The process can take up to 10 minutes per egg, but it's time well spent as it has been shown to deter predators. One of the key features of these beetles is that the head of the adults is hidden under their bulging pronotum (shoulder/thorax-covering).

INSECT inspector!

There are currently 19 species of Pot Beetle in the UK, however many have suffered declines in their distribution and are now quite rare.

CHESTNUT CLICK BEETLE

Anostirus castaneus
LENGTH 8–10 mm • HABITAT Open ground • RANGE Rare,
locations on Isle of Wight and North Yorkshire
COMPLETE METAMORPHOSIS

THE CHESTNUT CLICK BEETLE usually
lives on expanses of open ground, but
the larvae live in the soil and are largely
predatory, although they also eat the
roots of various plants. This distinctive
beetle has feathery antennae—rather like a
stag's antlers—and is an orangey brown/chestnut
colour. The area behind its head is covered in golden
hairs and it has a black tip at the end of its body.
Click beetles get their name because when threatened
or turned over, they jump into the air with a loud click.

CHESTNUT COLOUR

"FEATHERY" ANTENNAE

RHINOCEROS BEETLE

PRONOUNCED
"HORN"

Sinodendron cylindricum
LENGTH 15–18 mm • HABITAT Woods, parks and hedgerows
• RANGE Widespread from May to October • COMPLETE METAMORPHOSIS

THIS LARGE, CYLINDRICAL beetle
lives up to its name by sporting a
distinctive "horn" on the head of
the male. The female just has a
small bump, rather than a full
horn. The Rhinoceros Beetle is
glossy, purply-red or blue-black
and covered with small pits and
grooves. The adults are active in
the summer and are strong fliers,
although they can often be spotted
resting in the sun on dead tree trunks.
The larvae depend on rotten wood to live
in and feed on, while adults feed on tree sap.

PITTED WINGS

ROSEMARY BEETLE

Chrysolina americana
LENGTH 6–7 mm • **HABITAT Aromatic plants**
• **RANGE Widespread**
COMPLETE METAMORPHOSIS

ROSEMARY BEETLES EAT the foliage and flowers of aromatic plants such as Rosemary, Sage, Lavender and Thyme. Adult beetles are shiny with metallic purple and green stripes on their wing cases and thorax. The larvae are greyish-white with darker stripes running along their bodies and hatch from sausage-shaped eggs. The larvae feed from early autumn to spring. Adults can be found on plants throughout the year.

PURPLE AND GREEN STRIPED WINGS

....

RUGGED OIL BEETLE

SMALL, LUMPY FOREWINGS

Meloe rugosus **LENGTH 10–35 mm** • **HABITAT Flower-rich grassland on chalk, limestone and sandy soils** • **RANGE Central and southern England and Wales, from September through to April**
COMPLETE METAMORPHOSIS

OIL BEETLES GET their name from the toxic oily secretions they produce from their leg joints as a defence against predators. Females dig nest burrows in the ground to lay their eggs. Once the eggs hatch the larvae make their way onto flowers and wait for an unsuspecting bee to come along. The larvae use their hooked feet to grip onto the bee and catch a free ride to the bee's nest. The larvae then eat the bee's eggs and the store of pollen and nectar the bees have been working hard to collect for their nest. The larvae then stay in the bee's nest to develop until emerging as adults.

INSECT inspector!

In her two-month life cycle, a female Oil Beetle can lay up to 40,000 eggs.

SCARLET MALACHITE BEETLE

Malachius aeneus LENGTH **5–8 mm**
○ HABITAT **Meadows, thatched or timbered cottages**
○ RANGE **Extremely rare—only found on 8 sites in Essex, Cambridgeshire and Hampshire**
COMPLETE METAMORPHOSIS

THIS LITTLE BEETLE is one of the UK's rarest and most beautiful insects. The Scarlet Malachite Beetle is red and shimmering green. It is found in just eight sites in the UK. The adult beetles appear during April and May and feed on grasses and flowers in meadows, and often near thatched or timbered cottages during the summer months. The beetle lays its eggs in the thatch and timber. The Scarlet Malachite Beetle is a member of the soft-winged flower beetle family.

10s spotters

VIVID RED WINGS

RECTANGULAR BODY SHAPE

COMMON RED SOLDIER BEETLE

Rhagonycha fulva
LENGTH **10–15 mm** ○ HABITAT **Woodland margins, hedgerows and other rough places, often hunting on flowerheads**
○ RANGE **Widespread, from June to August**
COMPLETE METAMORPHOSIS

THE COMMON RED Soldier Beetle is also known as the "bloodsucker" for its striking red appearance—but it is harmless. It has a narrow, rectangular body and longish antennae and is bright orangey-red with black marks near the tips of its wing cases. During the summer, it is commonly found on open-structured flowers, such as daises and cow parsley. Adults feed on aphids, pollen and nectar. Larvae prey on slugs and snails, and live at the base of long grasses.

10s spotters

LONG ANTENNAE

INSECT inspector!

Soldier Beetles are so-named for their various combinations of black-and-red markings, which are reminiscent of a soldier's uniform.

STAG BEETLE

Lucanus cervus LENGTH 50–75 mm ▪ HABITAT Prefers Oak woodland, but can be found in gardens, hedgerows and parks ▪ RANGE Southeast England, particularly in South and West London, from May to August COMPLETE METAMORPHOSIS

THE STAG BEETLE is the UK's largest beetle. The male's antlers are overgrown jaws, used for wrestling contests if two males claim the same female. The female has normal jaws. Males have reddish-brown bodies. Females look like Lesser Stag beetles, but are larger, with smaller heads and brown wing cases instead of black ones. Larvae live in decaying wood, and can take up to six years to develop before they turn into adults. Adults have a much shorter life span: they emerge in May with the sole purpose of mating, and die in August once eggs have been laid.

FEARSOME JAWS

REDDISH-BROWN BODY AND WINGS

....

STREAKED BOMBARDIER BEETLE

LIGHT BROWN "STREAK"

Brachinus sclopeta
LENGTH Up to 7.5 mm ▪ HABITAT Piles of rubble and loose material, hiding in nooks and crannies ▪ RANGE Rare, East London and Margate COMPLETE METAMORPHOSIS

THE STREAKED BOMBARDIER Beetle possesses a remarkable but effective defence mechanism—it can spray a boiling, noxious chemical mixture from the tip of its flexible abdomen with an audible explosive sound. This boiling spray has the potential to kill other insects and strongly deter larger predators. Streaked Bombardiers have a metallic blue-green wing case with a narrow orange-red head and a distinctive red dash along its back—the Common Bombardier Beetle does not have this "streak".

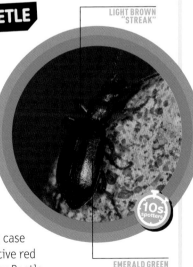

EMERALD GREEN WINGS

TANSY BEETLE

Chrysolina graminis
LENGTH 7.7–10.5 mm ◦ **HABITAT** Riverbanks and wetlands
◦ **RANGE** Restricted to the banks of the River Ouse, York,
North Yorkshire and a small colony in Cambridgeshire
COMPLETE METAMORPHOSIS

THE ENDANGERED TANSY

Beetle is a large and iridescent
Green Leaf Beetle with a
coppery sheen. The female
Tansy Beetle is generally
larger-bodied than the male.
They are specialist herbivores,
mainly eating the Tansy plant,
around which they complete their
entire life cycle. Adults are active from
April to June when they feed, mate and lay
eggs. Eggs hatch between May and July into larvae,
which feed hungrily on Tansy leaves.

SHINY GREEN HEAD,
TORSO AND WINGS

SHORT, STOCKY
BODY SHAPE

LONG ANTENNAE

LARGE UPPER LEGS

THICK-LEGGED FLOWER BEETLE

Oedemera nobilis
LENGTH 6–11 mm ◦ **HABITAT** Gardens, grassland, woods and coastal areas
◦ **RANGE** Widespread across the UK, common in southern and Southeast
England, from April to September
COMPLETE METAMORPHOSIS

THE THICK-LEGGED Flower Beetle
is often known by other names
including the Swollen-thighed
Beetle and the False Oil Beetle. It
is metallic green in colour with
thickened hind legs on males. It
is a pollinator of many open-
structured flowers including Cow
Parsley, Oxeye Daisy and Bramble.
These beetles are most frequently
spotted in bright sunlight on flower
heads on warm days.

VIOLET CLICK BEETLE

Limoniscus violaceus **LENGTH** 12 mm
• **HABITAT** Decaying wood, particularly Common Beech
and Ash • **RANGE** Very rare, found at just three sites in the UK
COMPLETE METAMORPHOSIS

THE VIOLET CLICK Beetle is an
extremely rare, elusive beetle
that only occurs in pasture
woodland on three sites in
the UK (one in Worcester-
shire, one on the Berkshire/
Surrey border, and one in
Gloucestershire). Very little is
known about the beetle, except
that it is found in the heart of
decaying trees, particularly
Common Beech and Ash, and breeds
in hollow trunks. It is a thin, black beetle,
with a blue sheen. The larvae, called
"wire-worms", are long, thin, whitish
grubs, resembling meal worms.

LONG BODY

10s spotters

INSECT inspector!

The Violet Click Beetle gets its
name from its habit of springing
upwards with an audible click if
it falls on its back.

VIOLET OIL BEETLE

SHINY HEAD
AND THORAX

Meloe violaceus **LENGTH** 10–33 mm • **HABITAT** Woodland edge habitats, glades
and rides, upland moorland and on flower-rich grassland • **RANGE** Rare
COMPLETE METAMORPHOSIS

THESE BEETLES HAVE a striking appearance
despite their black coloration—as light is
refracted off their lustrous shell, they
display a purple, blue or green sheen. When
they first emerge as adults, their abdomen
is small and compact but, as they gorge
themselves on soft grasses, their abdomen
becomes distended and can extend some
way beyond the tips of their wings. They can
often be found sunning themselves on paths and
females are sometimes seen digging burrows in
patches of bare ground to lay their eggs.

10s spotters

SMALL FOREWINGS

WASP BEETLE

Clytus arietis
LENGTH 16 mm · **HABITAT Woodland rides and hedgerows** · **RANGE Widespread in England and Wales, but rarer in Scotland, from May to July**
COMPLETE METAMORPHOSIS

THE WASP BEETLE is a small, narrow-bodied longhorn beetle with relatively short antennae. A clever mimic, it has black-and-yellow bands on the body and moves in a jerky fashion—fooling predators into thinking it is a more harmful Common Wasp. The larvae live in warm, dry, dead wood, such as fence posts and dead branches, and particularly favour Willow and Birch. Adults can be found feeding on flowers along hedgerows during the summer.

BLACK-AND-YELLOW MARKINGS

10s. spotters

LONG, BROWN LEGS

WOODWORM BEETLE

Anobium punctatum
LENGTH 2.7–4.5 mm · **HABITAT Damp wood** · **RANGE Common**
COMPLETE METAMORPHOSIS

THE MOST COMMON form of Woodworm Beetle is the Common Furniture Beetle. Its larva is a wood-boring grub that attacks softwood species of timber feeding on starch that is found in the wood. It generally prefers damp, rather than dry wood and the grub will head for, and stays in, plywood for longer than any other timber. The adult beetle has a brown ellipsoidal body with a pronotum resembling a monk's cowl.

INSECT inspector!

The adult Woodworm Beetle does not feed on wood but as it emerges from its pupa and exits the wood, it creates holes that are often the first sign of a woodworm infestation. The adult beetle only lives a matter of days, as its sole purpose is to mate and then lay eggs.

WORMWOOD MOONSHINER

Amara fusca
LENGTH 7.5–8.5 mm · HABITAT Heathland or
sand dunes · RANGE A very restricted
UK distribution
COMPLETE METAMORPHOSIS

REDDISH-BROWN
ANTENNAE

THE WORMWOOD MOONSHINER is a ground beetle that comes out at night to feed on Wormwood seeds. It varies from a pale bronzy brown to dark brown and lives on sparsely vegetated, dry, sandy or gravelly soil—such as on heathland or sand dunes. While the adults climb about in plants at night feeding on seeds, the larvae live on the ground and are predatory. The adults don't like just any seeds, they have a special liking for the seeds of a rare plant—the Breckland Wormwood.

10s
spotters

RIDGED WINGS

This beetle is unusual in that it is active even in extremely cold weather. Wormwood Moonshiners have been observed foraging for seeds when the temperature was down to -6o°C!

Laugh Out Loud!

Joke: Why did the entomologist*
get rid of his woodworm larva?

Answer: Because they were boring

*an entomologist is someone who studies insects

INSECT REPORT: THE ART OF DECEPTION

Hornet Hoverfly

Volucella zonaria

The Hornet Hoverfly looks like a dangerous stinging hornet but it is in fact harmless. This mimicry helps keep predators away (see page 45).

All living things need to protect themselves. Insects are no exception. Some insects do bite and sting potential predators, but others have mastered the art of deception. These insects rely on mimicry, camouflage, and deceptive behaviours to keep them safe .

YOUR BEST BET AT FINDING THIS SMOOTH STICK INSECT
(CLITARCHUS HOOKERII) IS SOUTHWEST ENGLAND IN THE
EARLY AUTUMN. THE UK DOESN'T HAVE ANY NATIVE
SPECIES OF STICK INSECT BUT THREE NEW ZEALAND
SPECIES HAVE BECOME SUCESSFULLY ESTABLISHED
HERE OVER THE LAST CENTURY.

Wasp Beetle

Clytus arietis

This beetle's black-and-yellow bands fool predators into
thinking it is a more harmful Common Wasp when they are
feeding on summer flowers. Wasp Beetles sometimes hatch
out of firewood that has been brought into the house to dry
over the winter (see page 34).

Scorpion Fly

Panorpa communis

Adult Scorpion Flies might
look a bit scary but they
are actually harmless! The
scorpion-like tail doesn't
sting, but is used by the
male fly in courtship
displays (see page 102).

Comma Butterfly

Polygonia c-album

Its underside has cryptic brown colouring, making it look
like a dead leaf. It gets its name from the comma-shaped
white spots on the underneath of its wings (see page 84).

COMMON EARWIG

Forficula auricularia
LENGTH 13–18 mm • HABITAT Dark, confined, damp areas such as under potted plants and in cracks between pavers • RANGE Throughout the UK, from January to December
INCOMPLETE METAMORPHOSIS

THICK PINCERS (MALE)

SHORT WINGS

THIS FLAT, REDDISH-BROWN insect has very short wings. It rarely flies and prefers to hitchhike in bundles of newspapers, luggage, cut flowers, and other objects. You can find them hanging out in high-moisture areas such as log piles, under plant pots or in damp sheds. At times, large numbers of these insects may seek shelter in and around homes. When they do, they don't damage property but they can be a huge nuisance because they invade everything.

Female Common Earwigs take parenthood to the extreme—for an insect. The mother grooms and turns her eggs, regularly shifting their position. Once the eggs hatch, she guards the nymphs and feeds them until they have moulted three times. She doesn't eat that whole time.

Common Earwigs have a strong pair of pincers, or cerci, on the end of their abdomen. They use their cerci to grab and hold onto prey. You can use them to separate the males from the females. On a male (like the one pictured above), the cerci are thick, curved, and separated quite far apart at the base. On a female, the cerci are thin, straight, and close together.

INSECT inspector!

Despite popular belief, and its name (from the Old English for "ear beetle"), the Common Earwig will not crawl into your ear while you sleep—it much prefers a nice log or stone pile! It feeds on organic matter, recycling important nutrients.

CRANEFLY

ELONGATED THORAX

Tipulidae
LENGTH 16 mm (body); 50 mm (leg)
- **HABITAT** Gardens and fields, often coming indoors
- **RANGE** Widespread, from June to September
COMPLETE METAMORPHOSIS

THE FAMILIAR DADDY longlegs is a large type of Cranefly, of which there are 94 species in the UK. An adult is brown, long-bodied, with translucent wings and very long legs (which easily fall off if handled). As a group, craneflies are unmistakable, although telling the different species apart can be difficult and often requires a microscope. As a larva, it is a grey grub that lives underground, feeding on stems and roots. This makes it unpopular with gardeners as it can leave bare patches of lawn.

LONG, THIN WINGS

BLACK HORSE FLY

Tabanus atratus
LENGTH 20–28 mm · **HABITAT** Damp pastures, especially near wood
· **RANGE** Common, from June to August
COMPLETE METAMORPHOSIS

BLACK HORSE FLIES ARE notorious biters of horses and humans and leave behind painful sores. Females feed on blood, their scissor-like mouth parts cut open flesh, allowing blood to ooze out. Males do not bite and do not drink blood. Instead, males drink flower nectar and spend their days looking for females to mate with. Males and females are both completely black, but males have huge eyes that touch each other at the center of the face; the eyes of females are separated.

LARGE EYES

BLACK HEAD, BODY AND LEGS

BLACK-WINGED BEE-FLY

Anthrax anthrax
LENGTH 10 mm • HABITAT Parks and gardens
around human habitation • RANGE Southern
England only, very rare
COMPLETE METAMORPHOSIS

THIS FLY IS COMMON
across mainland Europe
but very rarely seen in
the UK—only occasional
sightings have been
reported in the last
decade. The female lays its
eggs near the nests of bees
from where the fly larvae can feed
on the bee larvae! It is a large darkly
coloured fly with white marks on its
thorax.

HAIRY THORAX

DARK, OPAQUE
WINGS

BLUEBOTTLE

Calliphora vomitoria
LENGTH 10–12 mm • HABITAT Common in and around human
habitation • RANGE Common, from January to December,
often basking on walls in winter sunshine
COMPLETE METAMORPHOSIS

**THIS IS THE COMMON
BLOW-FLY** that makes lots of
noise as it buzzes around our
houses in search of a way
out. Females are strongly
attracted to meat or fish of
any kind, where they lay their
eggs. Males are more likely to
be found on flowers. The
Bluebottle's head and thorax
are dull grey, its abdomen is
bright metallic blue with black
markings and its body and legs are
covered with black bristle-like hair. Its
chest has spikes to protect it from other flies.

SHINY, BLUE
THORAX

LARGE, REDDISH-
BROWN EYES

DARK-EDGED BEE-FLY

Bombylius major
LENGTH **Up to 18 mm** • HABITAT **Gardens, parks and woodland** • RANGE **Widespread from April to May**
COMPLETE METAMORPHOSIS

THE DARK-EDGED BEE-FLY, or "Large Bee-fly", looks rather like a bumblebee, and buzzes like one too. It is our largest and most common bee-fly and uses its long, straight mouthpart to feed on nectar from primroses and violets during spring. It has yellowy-brown hair on its body, long, spindly legs, and its wings have dark markings along their leading edges, hence it's common name. The larvae of this insect are nest parasites of ground-nesting and solitary bees, feeding on the bee grubs.

FURRY BODY
LONG LEGS

FRUIT FLY

Drosophila Melanogaster
LENGTH **3 mm** • HABITAT **Overripe or rotting fruit** • RANGE **Common**
COMPLETE METAMORPHOSIS

FRUIT FLIES ARE A SMALL, common fly found near ripe or rotten fruit. They have red eyes, are yellow-brown in colour, and have transverse black rings across their abdomen. Males are slightly smaller than females and the back of their bodies is darker. Fruit Flies have a short life cycle of just two weeks. Their wings can beat at up to 220 times per second.

STRIPY THORAX

THIN, TRANSPARENT WINGS

GREENBOTTLE

Lucilia caesar
LENGTH 8–10 mm **HABITAT** Grassland and gardens
RANGE Common and widespread, from April to October
COMPLETE METAMORPHOSIS

THE GREENBOTTLE has brilliant, metallic, bright-green or bluish-green coloration with black markings. It develops a coppery ting with age. It has short, sparse black bristles and three cross-grooves on its thorax. Its wings are clear with light brown veins, and its legs and antennae are black. Adult flies typically feed on pollen and nectar of flowers. The larvae feed mainly on dead animals.

SHINY, GREEN
THORAX

Genetically, humans are about 44% similar to flies!

....

COMMON HOUSE-FLY

Musca domestica **LENGTH** 5–8 mm **HABITAT** Human habitation, especially around farms and rubbish dumps with abundant decaying matter **RANGE** Common, from January to December but most noticeable in summer
COMPLETE METAMORPHOSIS

LOOK FOR THE grey-and-black striped thorax and the yellow or orange abdominal patches. House-flies have red eyes, set farther apart in the slightly larger female. They are less common in houses than in the past because of better sanitation but their liking for excrement means that they carry numerous disease-causing germs. Adult flies usually live for two to four weeks, but can hibernate during the winter.

GOLDEN-
COLOURED
ABDOMEN

ST MARK'S FLY

Bibio marci LENGTH 12–14 mm · HABITAT Rough grassland, hedgerows, woodland margins and scrub · RANGE Widespread, from April to May
COMPLETE METAMORPHOSIS

HUGE SWARMS OF THESE jet-black flies drift over shrubs and other vegetation in spring, often around St Mark's Day (25 April). With their legs dangling in sinister fashion, the flies may cause alarm, but they are quite harmless. The female has much smaller eyes than the male. They breed in soil and rotting vegetation and are important pollinators of fruit trees and other plants.

BLACK THORAX

SPINDLY LEGS

10s spotters

Laugh Out Loud!

Joke: What do you call a fly without wings?

Answer: A walk!

YELLOW FLAT-FOOTED FLY

Agathomyia wankowiczii
LENGTH 3–6 mm · HABITAT Woodland · RANGE Rare
COMPLETE METAMORPHOSIS

THIS VIVID ORANGE FLY is the only invertebrate in the UK known to cause a gall on a fungus. A gall is an abnormal growth commonly found on plants but a gall caused by an insect on a fungus is very rare. The gall is found in large clusters on the whitish underside of the fungus. They start as small warts that can grow up to 10 mm in height. Inside each wart is the grub of the fly. Once the grub is fully grown, it bores a hole and falls to the ground where it buries itself into the soil before it pupates to turn into an adult fly.

ORANGE BODY

10s spotters

LENGTHWISE LINES ON WINGS

GOLDEN HOVERFLY

Callicera spinolae
LENGTH 10 mm ○ **HABITAT Ancient woodland and
parkland in Southeast England** ○ **RANGE Rare**
COMPLETE METAMORPHOSIS

LOOK OUT FOR THE Golden
Hoverfly on pollen and
nectar-rich ivy in
summer months.
Although it has only
been seen in four
locations in the last
ten years, it is possible
that it lives high in the
tree canopy and could be
under-recorded. Adults are
large and furry with long
black antennae that are white
at the tip. The larvae develop in
wet rot holes where they feed on
bacteria and other microbes, so the Golden
Hoverfly prefers old trees which are more likely to
produce the correct microhabitat.

BLACK EYES

**BLACK-TIPPED
LEGS**

10s
spotters

INSECT inspector!

Garden pests such as aphids can bring plant diseases and
damage crops, but hoverfly larvae are great at eating
them. This means actively encouraging hoverflies into
your garden is a great way to try to control the pests without the use of chemicals. One simple way to encourage
them is to plant the types of flowers hoverflies prefer,
such as Camomile, Parsley and Buckwheat. As an
added bonus, they also add some colour to the garden!

YELLOW FACE

BROWN THORAX WITH BLACK MARKS

HORNET HOVERFLY

Volucella zonaria
LENGTH 15–20 mm • HABITAT Parks and gardens
• RANGE Common in southern England, but spreading
north, from May to October
COMPLETE METAMORPHOSIS

THE HORNET HOVERFLY looks like a dangerous, stinging hornet but is, in fact, harmless. This mimicry helps to keep predators away. The larva of the Hornet Hoverfly can live happily in the nests of wasps without getting stung. The larva eats the debris and rubbish in the wasp nest and, in return, the wasps have a free cleaner! The Hornet Hoverfly can be identified by chestnut patches at the front of its abdomen. The male has larger eyes than the female.

MARMALADE HOVERFLY

Episyrphus balteatus
LENGTH 10 mm • HABITAT Gardens, hedgerows, parks and
woodland • RANGE Widespread, from January to December
COMPLETE METAMORPHOSIS

LOOK FOR THE PATTERN of broad orange and narrow black bands on the abdomen to identify this very common hoverfly. The colour of these broad bands gives it its common name. Hoverflies swarm over flowers like Tansy, Ragwort and Cow Parsley, collecting pollen and nectar and also feeding on honeydew. They often enter houses. The larvae eat aphids. The Marmalade Hoverfly can be seen in gardens, parks and sunny woodlands.

ORANGEY-YELLOW LEGS

PATTERNED ABDOMEN—
LONG STRIPES MIXED WITH
SHORTER STRIPES

COMMON MOSQUITO

Culex pipiens
LENGTH 6 mm ∘ HABITAT Small bodies of standing fresh water
∘ RANGE Almost everywhere, from January to December
COMPLETE METAMORPHOSIS

OUR MOST COMMON MOSQUITO, distinguished from most other brownish species by the white band at the front of each abdominal segment. In common with most of our mosquitoes, it rests with its body parallel to the feeding or resting surface. It can also be characterised by the presence of its sucking, elongated mouthpart. Females feed on vertebrate blood (including human blood) and males feed on nectar.

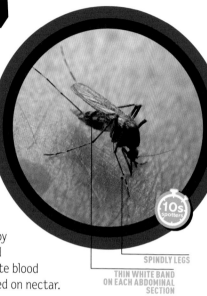

SPINDLY LEGS

THIN WHITE BAND
ON EACH ABDOMINAL
SECTION

WINTER GNAT

REFLECTIVE
WINGS

Trichocera annulata
LENGTH 8–10 mm ∘ HABITAT Woodland, parks and gardens
∘ RANGE Common, widespread, from January to December
but most noticeable in the colder months
COMPLETE METAMORPHOSIS

WINTER GNATS APPEAR during the colder months when the male gnats perform their courtship dances. Swarms gather in areas that are kept warmer by the sun. In low winter sunshine, the reflective wings of the dancing swarms can make them appear like apparitions— the gatherings are sometimes called "ghosts". Each male flies up and down to his own rhythm, but they cleverly space and pace themselves out to avoid colliding with others in the swarm.

LONG LEGS

HORNET ROBBERFLY

Asilus crabroniformis
LENGTH 20–30 mm
  HABITAT **Grassland, heathland and moorland**
  RANGE **Southern England and south Wales, from June to October**
COMPLETE METAMORPHOSIS

THE ROBUST, FAST-FLYING Hornet Robberfly is a predator, sitting and waiting on a suitable perch (such as a stone or pile of animal dung) for smaller insects to fly past, which it catches in midair. It prefers Dung Beetles, but will also eat bees and grasshoppers. Look for the stout beak, which penetrates its victims, the sturdy legs and the hairy face. The bristles protect the robberfly's eyes from its struggling victims. It has a brown thorax and a black-and-yellow abdomen.

LONG ABDOMEN, EXTENDS BEYOND WINGS

10s. spotters

HAIRY HEAD

INSECT inspector!

In Britain, there are nearly 7,200 species of fly—more than all the mammal species in the world added together! New fly species are being found all the time. We are used to seeing Bluebottles and House Flies because they often interact with people and where we live, but flies are found all over the planet. Some flies—like Fruit Flies—are tiny, while others—like Hornet Robberflies—are some of the UK's largest insects!

INSECT REPORT: MIGHTY MOUTHS

Bloody-nosed Beetle

Timarcha tenebricosa
This insect has quite the mighty mouth, and when threatened it oozes a distasteful red liquid to keep away any potential predators (see page 20).

March Brown Mayfly

Rhithrogena germanica
While there are many different types of mouth in the insect world, the adult mayfly has no functioning mouthparts at all! It does not feed throughout its entire adult life (see page 50).

Insects have an incredibly varied diet. Different foods can't be eaten in the same way, so the form and function of an insect's mouthparts need to match what it eats. Some insects—especially those that go through complete metamorphosis—change their diets as they move from one stage of development to the next, so their mouths have to change too.

THE ORANGE-TIP (*ANTHOCHARIS CARDAMINES* — SEE PAGE 88), IS A SMALL BUTTERFLY MOSTLY FOUND IN SPRING. THEIR CATERPILLARS ARE CANNIBALS, SO AFTER EATING THEIR OWN EGGSHELL, THEY MOVE ON TO OTHER ORANGE-TIP EGGS NEARBY.

Black Horse Fly

Tabanus atratus
These flies are notorious for biting horses and humans! However, it is the female that sucks blood, using her scissor-like mouth to cut open flesh. Males drink flower nectar, and spend a lot of their lives looking for mates (see page 39).

COMMON MAYFLY

Ephemera danica
LENGTH 15–30 mm
- **HABITAT Freshwater wetlands**
- **RANGE Widespread, from January to December**
INCOMPLETE METAMORPHOSIS

LOOK FOR THE DARK-SPOT-TED cream or greyish top of the abdomen to distinguish this insect from several similar species with spotted wings. Like all mayflies, the insects fly mainly by night. The name "mayfly" is misleading as many mayflies can be seen all year round. Adult mayflies are delicate creatures with broad, clear wings that have a lace-like appearance. They usually have short antennae and up to three very long, fine tail bristles. They hold their wings vertically, closed over their backs.

LACE-LIKE WINGS

TAIL BRISTLES

MARCH BROWN MAYFLY

Rhithrogena germanica
LENGTH 10–16 mm (body); 13–17 mm (wing length) • **HABITAT Large, clean rivers** • **RANGE Rare, from March to early April**
INCOMPLETE METAMORPHOSIS

THE MARCH BROWN IS PROBABLY the most famous of all British mayflies. Fishing flies, made of feathers, have been designed to look like them to help catch fish for over 500 years. Most mayflies like to emerge as adults during the summer months, however, the March Brown emerges right at the end of the winter. Adults leave the river around midday. The larvae drift in the water and then appear at the surface. They are an easy target for fish and birds so they emerge very quickly—often taking less than 30 seconds to moult and fly off the water!

BROWN PATTERNS ON WINGS

POND OLIVE

Cloeon dipterum

LENGTH 5–7 mm • HABITAT Rivers, streams and lakes
• RANGE Common and widespread, from May to October
INCOMPLETE METAMORPHOSIS

THE POND OLIVE is a reddish-brown mayfly. Males have extra eyes that are thought to enable them to locate isolated females in the mating swarm. Nymphs of this species live in pools and margins of rivers and streams or in ponds or shallow water in larger lakes. They swim in short, darting bursts and climb amongst the vegetation. Nymphs feed by scraping algae from submerged stones and other structures. Like all other species of mayfly in the UK, adults do not feed at all.

10s
spotter

TRANSPARENT
WINGS

INSECT
inspector!

While the mayfly nymph eats algae and some species eat other small insect nymphs, adults do not eat at all. In fact, mayfly adults have no functioning mouthparts. Like many other species of insect, once they reach adulthood their primary focus is on reproducing as quickly as they can, before they die.

BLACKFLY

Aphidoidea ssp.
LENGTH 2 mm • **HABITAT Found on a range of young plants**
• **RANGE Common and widespread, from April to August**
INCOMPLETE METAMORPHOSIS

BLACKFLIES ARE THE name for a group of aphids—tiny sap-eating bugs that are common across the country in springtime when they feed on the soft, younger parts of plants. Although they are very small, Blackflies often occur in huge numbers completely covering the parts of the plant they attack. As their name suggests, they are mainly black but with some white marks on their backs.

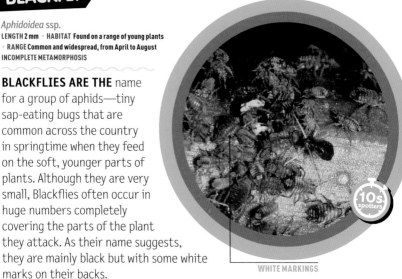

WHITE MARKINGS

GREENFLY

Aphidoidea ssp.
LENGTH 2 mm • **HABITAT Found on a range of young plants**
• **RANGE Common and widespread, from April to August**
INCOMPLETE METAMORPHOSIS

ANOTHER COMMON APHID group are Greenflies, which like their Blackfly cousins are sap-sucking insects that feed on young leaves and flower buds. Like many aphids, as they consume sap their bodies excrete a sticky liquid called honeydew which can lead to other problems for the plant by blocking light from its surface and promoting the growth of mould and fungus.

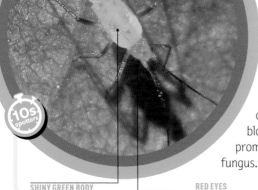

SHINY GREEN BODY   RED EYES

Pseudococcus calceolariae
LENGTH 4 mm · HABITAT Greenhouses, house plants · RANGE Common, from January to December
INCOMPLETE METAMORPHOSIS

MEALYBUGS ARE COMMON

insects that tend to live together in clusters in inaccessible parts of plants, such as leaf axils and under loose bark. Like aphids they suck sap from plants and then excrete the excess sugars as honeydew. The adult females have flattened oval-shaped bodies; they are sometimes pink in colour but appear whitish due to the white, waxy powder that covers their bodies.

WAXY COATED BODIES

WHITEFLY

Aleyrodidae ssp.
LENGTH 1–2 mm · HABITAT Vegetables, plants and trees · RANGE Common
COMPLETE METAMORPHOSIS

THESE INSECTS, with wings spanning no more than about 3 mm, look more like tiny moths than bugs. About 20 species occur in the UK, but they are not easy to distinguish. Nearly all feed voraciously on the underside of leaves, which become discoloured and distorted if the infestation is heavy. The insects and their wings are variously marked or mottled according to species, and many species are covered with fine wax powder, giving most species a floury, dusted appearance.

TINY WHITE BODY

COMMON BED BUG

Cimex lectularius
LENGTH 4–5 mm · HABITAT Dark locations, mostly close to where people sleep · RANGE Common
INCOMPLETE METAMORPHOSIS

THE COMMON BED BUG is light brown to reddish-brown, flat, oval, and has no hind wings. Bed Bugs are obligatory bloodsuckers with mouth parts that cut through the skin. The bite usually produces swelling with no red spot. They feed every five to seven days, which suggests that they do not spend most of their life searching for a host. When a Bed Bug is starved, it leaves its shelter to search for a host and then returns after successful feeding.

HORIZONTAL GROOVES ON ABDOMEN

NEW FOREST CICADA

Cicadetta montana
LENGTH 30 mm · HABITAT Lightly grazed, sunny, south-facing clearings and open sunny woodland rides and clearings bordered by scrub and tall trees · RANGE Extremely rare in UK, New Forest, Southern England, from late May to early July
INCOMPLETE METAMORPHOSIS

THE NEW FOREST CICADA is the only cicada native to the UK; however, these mysterious critters are highly endangered, and possibly extinct. Males are known for their characteristic high-pitched "song" which they produce by rapidly vibrating a tiny membrane on each side of their body. They perform this song to woo females for mating. The pitch is so high that it is at the limits of human hearing. The singing is only performed in still air conditions and in temperatures above 20°C.

SHORT ANTENNAE

LARGE WINGS EXTEND WELL BEYOND THORAX

FIREBUG

Pyrrhocoris apterus
LENGTH 8–12 mm ○ HABITAT Woods and most other well-vegetated places ○ RANGE Southern England, from January to December but usually dormant in winter
INCOMPLETE METAMORPHOSIS

THIS BRIGHTLY coloured bug usually always has reduced wings, leaving part of the abdomen exposed. Look for the completely black head and conspicuous round black spot on each forewing to distinguish this species. It feeds mainly on seeds of a mallow or lime tree but also attacks other insects. Adults hibernate in the soil and large swarms can be found on the ground in early spring.

BRIGHTLY PATTERNED BODY

BLACK HEAD, LEGS AND THORAX

10s spotters

INSECT inspector!

Firebugs—especially younger Firebugs—can often be seen forming large groups of tens or even hundreds of individuals, called aggregations. This behaviour seems to help the insect survive against some predators. It is not understood exactly what happens when a predator attacks an aggregation of bugs, but scientists have observed that after a bird has attacked a group it is less likely to attack even an individual Firebug in the future.

COMMON GREEN SHIELD BUG

Palomena prasina
LENGTH 13 mm ○ **HABITAT Gardens, farmland, grassland and coasts** ○ **RANGE Common, mainly found in England and Wales, but spreading north, from May to October**
INCOMPLETE METAMORPHOSIS

THE COMMON GREEN SHIELD BUG feeds on a wide variety of plants, helping to make this one species which could turn up anywhere from garden to farm. Adults overwinter and emerge in spring, laying their eggs on the undersides of leaves. The rounded nymphs appear in June and new adults are present in early autumn. The Common Green Shield Bug is bright green with tiny black dots and dark wings.

ROUNDED BODY SHAPE

SHORT STRIPES ON SHIELD

....

HAWTHORN SHIELD BUG

Acanthosoma haemorrhoidale **LENGTH 13–17 mm**
○ **HABITAT Gardens, parks and woodland** ○ **RANGE Widespread**
INCOMPLETE METAMORPHOSIS

THE RED-AND-GREEN Hawthorn Shield Bug is our largest shield bug. It feeds on Hawthorn, Rowan and Whitebeam. Look for the distinctly triangular shape and the bright green shield. The shield is surrounded by a bold red triangle formed by the forewings and the rear of the pronotum. The adults hibernate in moss and leaf litter over winter. It shares the name "stinkbug" with many other shield bugs because of the smelly fluid it emits when it's alarmed.

GREEN TRIANGLE

REDDISH-BROWN WINGS

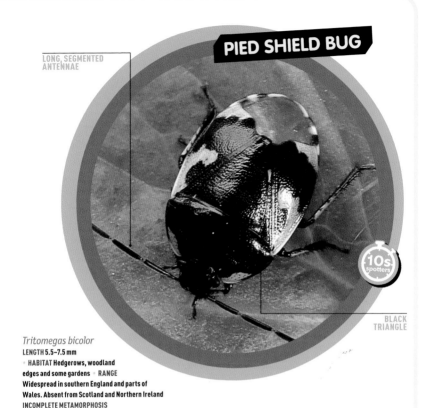

PIED SHIELD BUG

LONG, SEGMENTED
ANTENNAE

10s.
spotters

BLACK
TRIANGLE

Tritomegas bicolor
LENGTH 5.5–7.5 mm
HABITAT Hedgerows, woodland
edges and some gardens • **RANGE**
Widespread in southern England and parts of
Wales. Absent from Scotland and Northern Ireland
INCOMPLETE METAMORPHOSIS

THE PIED SHIELD BUG is often mistaken
for a black and white ladybird, especially the
young nymphs which have black spots but,
unlike these beetles, it has overlapping
wings, a very flattened body and piercing
mouthparts. It uses its mouthparts like a
straw to tap plant sap and has very specific
tastes, only supping from the seeds of
Dead-nettles and Black Horehound plants.
The young nymphs have orange-yellow
abdomens fading to yellowish-white as
they grow.

Laugh
Out Loud!

Joke: Which is the most
common season for
bedbug weddings?

Answer: The spring!

WESTERN CONIFER SHIELD BUG

Leptoglossus occidentalis
LENGTH 20 mm · HABITAT On and around conifer bark · RANGE Only recently found in UK, but becoming widespread in south of England and Wales
INCOMPLETE METAMORPHOSIS

THE WESTERN CONIFER

shield bug is very large and spectacular with a characteristic white zigzag mark across the centre of its forewings. It feeds on pines. It is attracted to light and may enter buildings in search of hibernation sites in the autumn. It makes a buzzing noise when airborne. Its primary defence is to spray a bitter, offending smell, though sometimes it can smell pleasantly of apples, bananas or pine sap.

BROWN BODY, LEGS AND HEAD

THIN HEAD

COMMON FROGHOPPER (OR CUCKOO SPIT INSECT)

Philaenus spumarius   LENGTH 5–7 mm · HABITAT Woodland, wetland and farmland · RANGE Common, from June to September
INCOMPLETE METAMORPHOSIS

IN SPRING AND SUMMER, look out for "cuckoo spit"—the frothy mass of bubbles that appear on plant stems everywhere. This is the protective covering for the nymphs of the tiny Common Froghopper. It protects the nymphs from drying out as well as from potential enemies. The adult pattern varies from black and white to many shades of brown; the nymph is green. The adult is a champion jumper and can leap 70 cm into the air—similar to a human jumping over a tower block!

GREEN OR BROWN BODY COLOURING

CHUNKY LEGS

RED AND BLACK FROGHOPPER

Cercopis vulnerata
LENGTH 8–12 mm • **HABITAT Most well-vegetated places, but most common in scrubby areas and on woodland margins** • **RANGE Widespread in England and Wales, from April to August**
INCOMPLETE METAMORPHOSIS

THIS BRIGHT AND unmistakable froghopper differs from others by spending its early life feeding on underground roots, although still surrounded by froth. It has an elongated and strongly shielded body. Red and Black Froghoppers are shiny black, with bright red marks on the wing-covers, one triangular mark at the base, one square mark in the middle and a stripe at the apex. These colours serve as a warning of their unpleasant taste. The hind wings are brownish, smoky and translucent.

STURDY BODY

WIDE, RED AND BLACK STRIPES

Froghoppers are amazing jumpers. The initial stages of the leap of the Common Froghopper are so powerful that a G-force of over 400 gravities is generated; in comparison, an astronaut rocketing into orbit experiences a G-force of 5 gravities.

POND SKATER

Gerris lacustris
LENGTH 10–12 mm
- **HABITAT** Ponds, ditches and quiet backwaters
- **RANGE** Common and widespread throughout the UK, from January to December but dormant in winter
INCOMPLETE METAMORPHOSIS

THE POND SKATER

has a thin, brownish-grey body and a small head with large eyes. It is usually seen skimming over the surface of still and slow-moving water. Little dimples show where its legs rest on the surface film. Its short front legs are used to grab other insects falling on to the water surface. Adults hibernate in debris on land, sometimes far from the water.

10s.
spotters

TWO SETS OF LONG BACK LEGS

SHORT FORELEGS

INSECT inspector!

Pond Skaters and other insects that appear to walk on water—such as Water Measurers and Water Boatmen—have hydrophobic (water fearing) hairs on their undersides or on their legs. These hairs actually repel the water and so prevent the insect from sinking below the surface.

SPINDLY
LEGS

COMMON WATER MEASURER

Hydrometra stagnorum
LENGTH 9–12 mm · HABITAT Ditches, ponds and slow-
moving streams · RANGE Fairly common, from January
to December but dormant in winter
INCOMPLETE METAMORPHOSIS

THIS NORMALLY WINGLESS
insect walks very slowly over the
surface of ponds and streams
and usually keeps close to the
marginal vegetation. It spears
mosquito larvae, water fleas and
other small prey by jabbing its beak
down through the water surface. It has
a very long, thin head, a slender body and
its antennae resemble a fourth pair of legs. It
is usually wingless but occasionally fully winged. It often has a
dark greyish or even black appearance.

LONG, THIN HEAD

OAR-SHAPED
LEG TIPS

LESSER WATER BOATMAN

Corixa punctata LENGTH 12–14 mm · HABITAT Weedy ponds, lakes
and ditches · RANGE Common, from January to December
INCOMPLETE METAMORPHOSIS

LESSER WATER BOATMEN feed by
sucking up debris from the bottom of
the pond. They fly well and the males
"sing" loudly by rubbing their front
legs against their heads. Lesser
water boatmen are dark brown with
yellow, close-set stripes. They have
long, oar-like legs to help them swim
at the surface of the water. Like other
aquatic bugs, they need to breathe at
the water's surface, but they have
developed an ingenious trick to allow them to
remain under water for longer: they hang upside-
down, collecting air from the water's surface and then carry it
around as a bubble on their body.

SHORT, STOCKY
BODY

INSECT REPORT: BITES & STINGS

Devil's Coach Horse

Staphylinus olens
Known for curling up it's tail like a scorpion when threatened, and letting off a foul-smelling substance—this insect is an aggressive predator that can deliver a painful bite to us humans (see page 23).

THE GERMAN WASP *(VESPULA GERMANICA)* IS ONE OF THE MOST COMMON WASPS IN THE UK. IT EATS OTHER INSECTS— USUALLY FLIES AND CATERPILLARS—AS WELL AS RIPE FRUITS AND NECTAR. UNLIKE MANY WASP SPECIES, GERMAN WASPS CAN DELIVER MORE THAN ONE STING AS THEY DO NOT DIE AFTER PASSING ON THEIR VENOM (SEE PAGE 79).

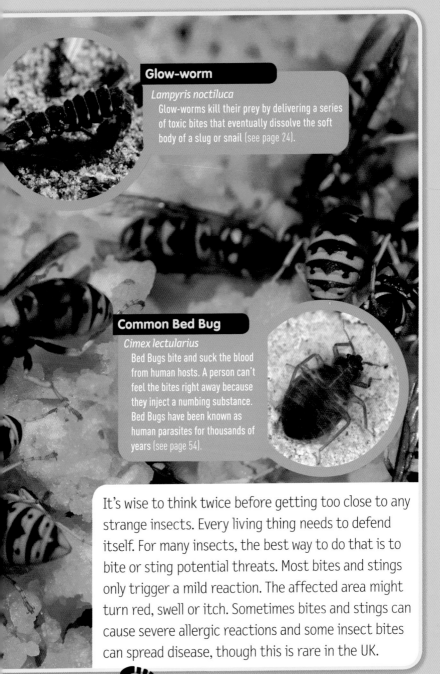

Glow-worm

Lampyris noctiluca
Glow-worms kill their prey by delivering a series of toxic bites that eventually dissolve the soft body of a slug or snail (see page 24).

Common Bed Bug

Cimex lectularius
Bed Bugs bite and suck the blood from human hosts. A person can't feel the bites right away because they inject a numbing substance. Bed Bugs have been known as human parasites for thousands of years (see page 54).

It's wise to think twice before getting too close to any strange insects. Every living thing needs to defend itself. For many insects, the best way to do that is to bite or sting potential threats. Most bites and stings only trigger a mild reaction. The affected area might turn red, swell or itch. Sometimes bites and stings can cause severe allergic reactions and some insect bites can spread disease, though this is rare in the UK.

BLACK LAWN ANT

Lasius niger

LENGTH 5 mm ○ **HABITAT** Nests in gardens, in lawns or flowerbeds, often under rocks and stones ○ **RANGE** Common and widespread across UK
COMPLETE METAMORPHOSIS

BLACK LAWN OR GARDEN Ants are one of the commonest species of ants found across Europe. As the name suggests they are usually black but dark brown individuals are not uncommon. Lawn Ants are found in large colonies containing thousands of individuals. They are aggressive and you can often see them attacking other insects when disturbed.

SHINY, BLACK HEAD AND THORAX

COMMON RED ANT

Myrmica rubra

LENGTH 6 mm ○ **HABITAT** Gardens and parks ○ **RANGE** Common, from January to December
COMPLETE METAMORPHOSIS

THESE VERY COMMON garden ants, called Red Ants, are actually pale brown. They have a two-seg-mented waist; both segments having a smoothly rounded hump. They nest under logs, stones and tree stumps, forming colonies that rarely contain more than 100 individuals. In common with all ants with a two-segmented waist, they have a sting.

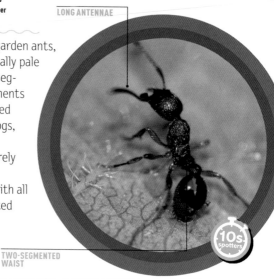

LONG ANTENNAE

TWO-SEGMENTED WAIST

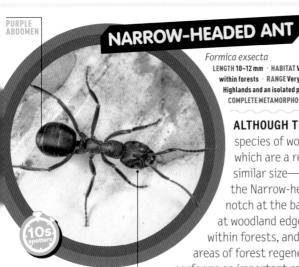

PURPLE
ABDOMEN

NARROW-HEADED ANT

Formica exsecta
LENGTH 10–12 mm · HABITAT Woodland edges and open areas
within forests · RANGE Very rare, restricted to Scottish
Highlands and an isolated population in Devon
COMPLETE METAMORPHOSIS

ALTHOUGH THERE ARE OTHER
species of wood ants—all of
which are a reddish colour and
similar size—you can identify
the Narrow-headed Ant by the deep
notch at the back of its head. It lives
at woodland edges and open areas
within forests, and is associated with
areas of forest regeneration. This ant
performs an important role in this regeneration
process, helping establish plants whose seeds are
dependent upon ants, such as the rare Small Cow-wheat.

THIN HEAD

WOOD ANT

Formica rufa
LENGTH 10–12 mm · HABITAT Woodland, building large nest mounds with
leaves and other debris · RANGE Widespread but localised throughout
England and Wales, from January to December but dormant in winter
COMPLETE METAMORPHOSIS

ONE OF SEVERAL large
mound-building ants. Look for the
largely black head and abdomen,
and the reddish-brown thorax.
The head is hairless. Note the
leaf-like scale, characteristically
between the thorax and the
abdomen. The ants do not sting,
but fire formic acid from their rear
ends when disturbed. They feed mainly
on other insects. Colonies contain up to
500,000 insects and are found mainly in
woodlands.

DARK RED OR
BLACK ABDOMEN

BUFF-TAILED BUMBLEBEE

Bombus terrestris
LENGTH 20–22 mm • **HABITAT Gardens and parks**
• **RANGE Common in lowland areas throughout the UK and Ireland from March to August**
COMPLETE METAMORPHOSIS

OUR LARGEST BUMBLEBEE, the Buff-tailed Bumblebee is named after the queen bee's buff-coloured tail. The worker bees have white tails with a faint buff line separating them from the rest of the abdomen. The Buff-tailed Bumblebee visits many different types of flowers for pollen and nectar; it has a short tongue, however, so prefers open, daisy-like flowers. Look for the orange or golden collar and second abdominal segment. This bumblebee nests underground in large colonies of up to 600 bees, often using the old nests of small mammals.

BUFF "TAIL"

INSECT inspector!

Buff-tailed Bumblebees are known as "nectar robbers": if they come across a flower that is too deep for their tongue, they bite a hole at its base and suck out the nectar.

COMMON CARDER BEE

ORANGE FURRED THORAX

Bombus pascuorum
LENGTH 13 mm • **HABITAT Gardens, farmland, woodland, hedgerows and heaths** • **RANGE Widespread, from March to September**
COMPLETE METAMORPHOSIS

THE COMMON CARDER BEE is a fluffy, brown-and-orange bumblebee, sometimes displaying darker bands on the abdomen. It is one of our most common bumblebees and nests in cavities, such as old mouse runs and birds' nests. It is a social insect; nests may contain up to 200 workers. Common Carder Bees are one of several "long-tongued bees" that feed on flowers with long tubular florets, such as Heather, Clover and Lavender.

5 OR MORE ORANGE STRIPES ON THE ABDOMEN

LARGE GARDEN BUMBLEBEE

Bombus ruderatus
LENGTH 15–22 mm · HABITAT Meadowland · RANGE It is now mainly found in the Fens, East Midlands and Cambridgeshire
COMPLETE METAMORPHOSIS

A LARGE BUMBLEBEE WITH a very long tongue, which is often held outstretched as the bee approaches a flower. It has a long face and tongue, which is well adapted for feeding on long-tubed flowers. Usually its body is black with two yellow bands and a single thin yellow band on the abdomen, but it can be completely black, making it one of the hardest bees to positively identify. Queens are the most difficult to find as the bee is scarce in many areas and queens are less abundant within a colony itself.

LESS OBVIOUS STRIPES ON ABDOMEN

....

LARGE RED-TAILED BUMBLEBEE

Bombus lapidarius
LENGTH 22 mm · HABITAT Gardens, farmland, woodland edges, hedgerows and heathland · RANGE Common from April to November
COMPLETE METAMORPHOSIS

THE FEMALE RED-TAILED Bumblebee is a very large, black bumblebee with a big red "tail". Males are smaller and, as well as the red tail, have two yellow bands on the thorax and one at the base of the abdomen. They nest underground, often in old burrows, under stones, or at the base of old walls. The Red-tailed Bumblebee is a very common bumblebee, emerging early in the spring and feeding on flowers right through to the autumn.

ORANGEY-RED "TAIL"

LACK OF DISTINCT STRIPES ON THORAX

SHRILL CARDER BEE

Bombus sylvarum
LENGTH 10–18 mm · HABITAT Grassland, farmland, coasts, wetlands, towns and gardens · RANGE Only found in a handful of locations in the UK, including large military ranges and unimproved pastures in Somerset, Gwent, Pembrokeshire, Glamorgan, and along the Thames corridor.
COMPLETE METAMORPHOSIS

THE SHRILL CARDER BEE is a small bumblebee. It has distinctive markings: it is grey-green in colour, with a single black band across the thorax, two dark bands on the abdomen, and a pale orange tip to the abdomen. The Shrill Carder Bee gets its name from the "shrill" buzz that it makes, which is higher in pitch than that of other bees. The queens tend to produce a higher-pitched buzz than the males and workers.

ORANGE TIP TO ABDOMEN

TAN AND BLACK FUR

TREE BUMBLEBEE

Bombus hypnorum
LENGTH 10–16 mm · HABITAT Woodland, towns and gardens · RANGE Found in England and Wales, and southern Scotland, from March to July
COMPLETE METAMORPHOSIS

THE TREE BUMBLEBEE HAS fuzzy, browny-orange hairs on its thorax, a black abdomen, and a white tail. It is associated with open woodland, so is commonly found in gardens that have a similar type of habitat. It nests in cavities, such as old birds' nests, bird boxes, or roof spaces. It visits a wide range of flowers, particularly those of soft fruits, such as blackberries.

ORANGE TUFT ON THORAX

WHITE "TAIL"

Tree Bumblebees can sometimes be found nesting low down in old mouse nests or even in the fluff of tumble drier vent pipes!

CUCKOO BUMBLEBEE

Bombus bohemicus
LENGTH 15–20 mm • **HABITAT Woodland,
towns and gardens, wherever host bees
are found** • **RANGE Across the UK,
especially northern England and
southern Scotland**
COMPLETE METAMORPHOSIS

CUCKOO BEE is a general name used to describe a large number of bees that display the same characteristic as the bird—that of laying their eggs in other bees' nests. Because they leave their eggs to be reared by other bees, they do not have pollen-carrying features, but usually the adult Cuckoo Bee will feed on pollen itself. **CUCKOO BUMBLEBEES** are more aggressive in their behaviour than most Cuckoo Bees such as the Six-banded Nomad Bee (see page 72). Instead of simply laying their eggs in a host nest, Cuckoo Bumblebees will invade the nest and kill the host queen. Very quickly they take over the colony eating host bee eggs and replacing them with their own.

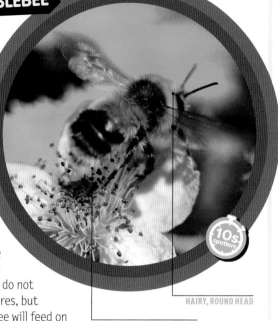

HAIRY, ROUND HEAD

PALE YELLOW, OR
WHITE "TAIL"

Laugh Out Loud!

Joke: A man walked into a pet shop and asked for
12 bees. The shop assistant came back with
a bag with 13 bees in it.
"You've given me too many" said the customer.
The shop assistant pointed to one of the bees
and said "That's a freebie!"

BLOOD BEE

Sphecodes ssp.
LENGTH 10–12 mm · **HABITAT Wide-ranging, depending
on availability of the hosts** · **RANGE Southern
England and Wales, April to September**
COMPLETE METAMORPHOSIS

BLOOD BEES ARE kleptoparasitic,
which means they steal nutrients
from other animals—in this case
other bees—to feed their young.
They get their name from the
dark-red or almost black colour of
their abdomen. Due to their nature,
they can be found anywhere near
their hosts (the bees they steal
nutrients from).

BLACK, FURLESS
THORAX

BLOOD RED ABDOMEN

IVY BEE

Colletes hederae
LENGTH 10–13 mm · **HABITAT Coasts, towns and gardens**
· **RANGE Found in southern England and Wales, and the
Channel Islands from September to November**
COMPLETE METAMORPHOSIS

THE IVY BEE LOOKS like a honey
bee; it has an orangey-brown,
hairy thorax, and distinct black
and yellow stripes on its
abdomen. Ivy Bees feed mainly
on Ivy, so they emerge to fit in
with the flowering period of
this plant: late September to
November. Ivy Bees nest in
loose, light or sandy soil on
southern-facing banks and cliffs
with Ivy nearby for foraging. They are
solitary bees, but when conditions are
suitable, there may be thousands of nests
in the same area.

STRIPED
ABDOMEN

ORANGEY-GOLDEN
FUR ON FACE

PATCHWORK LEAF-CUTTER BEE

Megachile centuncularis
LENGTH 13 mm ▪ HABITAT Grassland, woodland, farmland, towns
and gardens ▪ RANGE Common, from April to August
COMPLETE METAMORPHOSIS

THE PATCHWORK LEAF-CUTTER
bee looks like a dark Honey Bee,
but the underside of its
abdomen is orange. It is best
recognised by its habit of
carrying pieces of leaf back to
its nest. Leaf-cutter Bees nest in
holes in plant stems, dead wood,
cliffs or old walls, and can be seen
in gardens. They famously cut discs
out of leaves (they particularly like
roses), gluing them together with saliva
to build the "cells" in which their larvae live.

ORANGE UNDERSIDE

In the last hundred years bee numbers have declined around the
world. In Britain and Ireland, 13 bee species have become extinct in
that time and well over 30 more are considered to be in danger right
now. This is a major concern as bees play a crucial role in pollinating
the plants around our countryside—more than three-quarters of our
wild flowers are pollinated by bees. If the bees weren't around to pollinate these plants, many of the foods we enjoy eating as humans
would not grow so easily.

We can all help the bees by doing some simple things. If we plant
more bee-friendly flowers we will make it easier for bees to find the
nectar they need to thrive. We can also encourage our friends, family
and schools to plant wild flower gardens—they are colourful, look and
smell great, and provide interesting spaces in which to spend time. If
you don't have a garden you can plant bee-friendly flowers—like
Foxgloves and Daffodils—in a window box.

RED MASON BEE

Osmia bicornis
LENGTH 6–11 mm • HABITAT Cliffs and old buildings
• RANGE Widespread in England and Wales,
rarer in Scotland, from March to June
COMPLETE METAMORPHOSIS

THE RED MASON BEE is
covered in dense gingery
hair; the males are smaller
than the females and sport a
white tuft of hair on the face.
These small bees nest in hollow
plant stems, in holes in cliffs,
and in the crumbling mortar of
old buildings. They are solitary
creatures so, after mating, each
female builds its own nest; she lines
each "cell" with mud and pollen and lays a
single egg in each until the cavity is full. The
larvae hatch and develop, pupating in autumn and
hibernating over winter.

VERY SMALL
BODY SIZE

ORANGEY-
GINGER FUR

SIX-BANDED NOMAD BEE

Nomada sexfasciata LENGTH 10–12 mm
• HABITAT Soft cliffs • RANGE Rare, Prawle Point, south Devon
COMPLETE METAMORPHOSIS

AS THEIR NAME SUGGESTS Nomad Bees
have a nomadic or wandering lifestyle.
This is because they are a species of
"Cuckoo Bee" (see page 69), that has
to stay near its host so that it has a
place to lay its eggs. The Six-banded
Nomad Bee's host of choice is the
Long-horned Bee. The Nomad Bee's
lifestyle may seem to harm its host,
but the relationship is not one-sided.
The two species have evolved together
over millions of years. Long-horned bees
can cope with the Nomads, but the Nomads
need strong host numbers to survive.

YELLOW
BANDS

TAWNY MINING BEE

Andrena fulva
LENGTH 12 mm • HABITAT Grassland, farmland, heathland, moorland, towns and parks
• RANGE Widespread in England and Wales, but rarer in Scotland, from April to May
COMPLETE METAMORPHOSIS

THE TAWNY MINING BEE is a common, solitary bee that nests underground, building a little volcano-like mound of soil around the mouth of its burrow. The female is immediately identifiable by the black head and bright orange abdominal hairs. It often feeds at the flowers of gooseberries and currants. Sometimes called the lawn bee, it often nests in lawns and throws up little volcano-like heaps of soil around its burrow. Males are more slender and essentially black with scattered white tufts of hair on their face.

BRIGHT ORANGE
ABDOMINAL FUR
(FEMALE)

BURROWING
BEHAVIOUR

INSECT inspector!

When asked to think of a bee, most people think of a Honey Bee collecting pollen to take back to the hive to use as a food supply over the winter. Not all bees live like this. There are many types of bees that do not live in hives or big nests at all—they are called solitary bees because they only live with and care for their own offspring. Leaf-cutter Bees and Mining Bees are solitary bees, as are many of the parasitic bees like the Blood Bees and Cuckoo Bees. Solitary bees do not make honey and they do not have queens.

GIANT WOOD WASP

Urocerus gigas
LENGTH 40 mm ○ HABITAT In, or close to,
coniferous woods ○ RANGE May to October
COMPLETE METAMORPHOSIS

THE GIANT WOOD WASP
looks similar to a hornet,
but it's bigger and is
actually a sawfly, a wasp
relative. It is black and
yellow and is sometimes
referred to as a "Giant
Horntail" due to the female's
long, pointed tube at the back
of her body. This tube is often
mistaken for a sting but, in fact, it
is used to lay eggs in the trunks of
coniferous trees. Despite its rather
threatening appearance, the Giant Wood Wasp is
completely harmless.

10s
spotters

ORANGE
ABDOMEN

BLACK
THORAX

INSECT inspector!

Sawflies do not sting, but their appearance—resembling wasps—
may put many predators off trying to eat them. There are some
predators however, that do not seem to mind eating even stinging
insects. Badgers will often tackle a wasp nest to get to the tasty
larvae inside and many birds—in particular Starlings and
Magpies—will eat individual insects if they can find them.

Giant Wood Wasps are particularly vulnerable to attack
from birds because the process of digging into tree bark
and laying their eggs takes a long time. If the eggs are
successfully laid, larvae will live in the tree for
several years before emerging in winged adult
form.

TRANSPARENT WINGS

WILLOW SAWFLY

10s spotters

BLACK HEAD, THORAX AND ABDOMEN

Nematus oligospilus

LENGTH 8–10 mm
- **HABITAT** Willow trees
- **RANGE** Late spring to early autumn

COMPLETE METAMORPHOSIS

WILLOW SAWFLIES ARE slender insects with wings that are transparent and have black veins. They have a pair of long antennae and black eyes. Willow Sawflies lay their eggs in Willow leaves and the developing larvae feed on the leaves. They then either fall off the tree and form a cocoon in the soil under it, or they form a cocoon on the tree. The Sawfly pupates inside the cocoon, in the spring, and changes into an adult.

Laugh Out Loud!

Joke: A boy walks into a pet shop and asks the shop assistant, "Can I have a wasp please?"
The shop assistant says, "Don't be silly, we don't sell wasps."
The boy asks, "Well why do you have one in your window then?"

ASIAN HORNET

Vespa velutina
LENGTH 30 mm • HABITAT Grassy or lightly wooded areas near Honey Bees
• RANGE South of England, particularly near the coast and in the Channel Islands
COMPLETE METAMORPHOSIS

THE ASIAN HORNET IS a Far Eastern and South Asian species accidentally introduced into France from China and now spreading over the European mainland. It has been in the UK since 2016. It is a pest of Honey Bee hives as it evolved in areas where Honey Bees occur in the wild and has a natural liking for them—more so than our native hornet. Asian Hornets can sting but the sting is no more painful than that of any other species found in the UK or Ireland.

DARK BROWN
THORAX
ORANGE
STRIPES

EUROPEAN HORNET

YELLOW STRIPED
ABDOMEN

Vespa crabro
LENGTH 25–35 mm • HABITAT Hollow trees and sheltered cavities
• RANGE Widespread in southern and central UK and spreading
further north, best seen May to November
COMPLETE METAMORPHOSIS

A VERY LARGE RELATIVE of the Common Wasp, the Hornet lives mostly in woodland, parkland and gardens. Queen Hornets emerge from hibernation in spring and start to build their nests by chewing up wood—these "paper" nests are often made in hollow trees, or in cavities in buildings. Inside the nest, sterile workers hatch and look after the new young produced by the queen. At the end of summer, reproductive males and queens develop and leave the nest to mate. The males and previous queen die, and the new females hibernate, ready to emerge next spring and start the cycle again. Hornets catch a wide variety of invertebrates, mainly to feed to their larvae; they feed themselves on high-energy substances like nectar and sap.

Nysson trimaculatus
LENGTH 12 mm · HABITAT Brownfield sites,
bare open ground and rich grassland
· RANGE Common, found all over the
country in the summer months
COMPLETE METAMORPHOSIS

10s.
spotters

USUALLY MORE
BLACK AND LESS
YELLOW STRIPES

DIGGER WASPS are fairly common in the UK. As the name suggests, they burrow or "dig" into the ground when nesting. Digger Wasps are a type of solitary wasp meaning that females make a nest for their own young. This nesting behaviour is different to social wasps, as female social wasps co-operate with their sisters and their mother in the maintenance of a colony that may well contain hundreds or even thousands of workers, as well as a queen. Digger Wasps are hard to tell apart from social wasps although when resting they do not fold their wings lengthwise as other wasps do.

INSECT
inspector!

Wasp Fact: No male wasps or male hornets are able to sting and are therefore safe to approach and even pick up. It is the same with male bees. However it can be really difficult to tell the difference between a male and female bee or wasp, so we would always caution against getting too close!

COMMON WASP

Vespa vulgaris
LENGTH 12–17 mm (the queen is 20 mm)
◦ **HABITAT** Gardens, meadows, woodland, and urban areas
◦ **RANGE** From late spring to late summer
COMPLETE METAMORPHOSIS

THE COMMON WASP'S iconic black and yellow stripes warn other creatures that they are dangerous. With the abdomen split into six segments, one black/yellow stripe on each, the Common Wasp is similar to the German Wasp. To build a nest, a Common Wasp queen creates a cylindrical column (petiole), then a single cell which she surrounds with a further six cells, giving the cells their characteristic hexagonal shape. She builds cells in a layer until she has 20–30 then lays an egg in each. Each Common Wasp nest may contain between 5,000 and 10,000 individuals.

CLEAN YELLOW LINE BEHIND EYE

10s spotters

DOTS ON ABDOMEN, CLEAR OF BLACK STRIPES

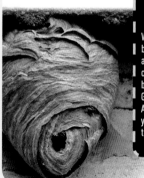

DANGER! ☠

Wasps are one of the insects we humans fear the most and with good reason because a wasp sting can be very painful! In some rare cases stings can cause an allergic reaction which can be fatal—although this is in a tiny number of cases, usually less than 5 per year in the UK. What can be more dangerous is being stung multiple times. Many wasps—including the Common Wasp and German Wasp do not die after stinging and so can sting a number of times. Also, most UK wasps live in nests—so sometimes when you disturb one wasp you are disturbing lots! **NEVER** approach a wasps' nest. If you see one, report it to an adult so it can be removed by professionals.

GERMAN WASP

Vespula germanica
LENGTH 12–20 mm · **HABITAT Woodland, hedgerows and other rough places** · **RANGE From April to October**
COMPLETE METAMORPHOSIS

THE GERMAN WASP is one of the most common wasps found in the UK. It eats other insects, usually flies and caterpillars, as well as ripe fruits and nectar. The German Wasp's nest is like that of the Common Wasp but greyish in colour and often made underground. Building the nest usually starts in April. The nests can get very big with thousands of wasps living there, and may have more than one way in. If their nests are disturbed, German Wasps become very aggressive and are likely to sting.

DOTS ON ABDOMEN OFTEN MERGE WITH STRIPES

5-BANDED WEEVIL WASP

Cerceris quinquefasciata
LENGTH 10 mm · **HABITAT Southern England, in areas of bare sand exposed to the sun** · **RANGE From July to August**
COMPLETE METAMORPHOSIS

THE 5-BANDED WEEVIL Wasp is a small yellow and black solitary wasp. This means it lives on its own rather than in a colony. It catches Weevils by paralysing them with its sting, then keeps them in its nest. It lays an egg into the Weevil, which then hatches and eats it! Although it is quite widely distributed in the south of England (especially in the southeast), the 5-Banded Weevil Wasp is considered a rare species.

NARROW BODY SHAPE

RUBY-TAILED WASP

GREEN-BLUE HEAD AND THORAX

RED ABDOMEN

Chrysis ruddii
LENGTH Up to 10 mm · HABITAT Dry meadows, forest margins and sun-exposed rock cliffs · RANGE From April to September
COMPLETE METAMORPHOSIS

RUBY-TAILED WASPS are solitary, meaning they don't live in large social nests. They are parasites, so they need another species for part of their life cycle. A female Ruby-tailed Wasp finds the nest of its host insect, reverses into the host's nest hole and lays its eggs next to the host eggs. The wasp eggs hatch into larvae, which eat the newborn host species. The unsuspecting adult host returns to seal its nest hole, never knowing that the Ruby-tailed Wasp is inside! The larvae complete their development inside the nest and the adults emerge the following spring.

INSECT Inspector!

The Ruby-tailed Wasp has a very hard body cuticle that protects it from stings from other insects. If it really feels it is in danger this wasp can curl up into a ball.

SABRE WASP

BLACK-AND-WHITE-STRIPED ABDOMEN

Rhyssa persuasoria LENGTH 20–100 mm · HABITAT Coniferous woodland · RANGE Widespread but localised, peaking in July and August
COMPLETE METAMORPHOSIS

YOU MAY SPOT THE FEMALE Sabre Wasp tapping on wood piles in coniferous woods with its antennae in search of extremely well-hidden insect larvae. She then uses her long egg-laying tail to drill a "probe" hole. She may drill a few probe holes before deciding on an appropriate position, and then drills as deep as she can. After 30–60 minutes, if successful, she will breach the tunnel wall, sting the larvae and then lay an egg on its body. With the larvae paralysed by the sting, it awaits the inevitable demise of being consumed alive by the Sabre Wasp grub.

LONG TAIL

YELLOW ICHNEUMON

ORANGE (NOT YELLOW) HEAD, THORAX AND ABDOMEN

Enicospilus ramidulus

LENGTH 20 mm • **HABITAT Woodland. Nocturnal, though the males can be found flying by day**
• **RANGE Widespread. Their hosts are moth caterpillars especially owlet moths**
COMPLETE METAMORPHOSIS

ICHNEUMON WASPS are parasitoids—creatures that live in very close contact with a host. In the case of Ichneumon Wasps, the adult female inserts her eggs inside the body of a caterpillar and the Ichneumon larva devours the caterpillar from the inside—usually killing it in the process. Ichneumon Wasps are solitary wasps with narrow waists and very long antennae. The female has an ovipositor at the tip of its abdomen instead of a stinger.

INSECT REPORT:
JUST WING IT!

THE LARGE WHITE *PIERIS BRASSICAE* IS THE LARGEST WHITE
BUTTERFLY IN THE UK. BUTTERFLY WINGS ARE COVERED IN
FINELY TEXTURED, TINY SCALES WHICH CREATE THE INTRICATE
PATTERNS WE SEE. OFTEN THE UNDERSIDE OF THE WING
LOOKS QUITE DIFFERENT FROM THE UPPERSIDE.

Insects are the only invertebrates that can fly. How
well an insect flies depends on the type of wings it
has as they come in all shapes and sizes! Some are
thick, others thin. They can be leathery, scaly, or
smooth. Some wings are used for flying, others used
for protection, collecting heat, or even attracting
mates. Although most adult insects have two pairs
of wings, some only have a single pair.

Banded Demoiselle

Calopteryx splendens
One of only two British damselfly species with coloured wings—the other is the Beautiful Demoiselle—the male Banded Demoiselle has a large, dark smudge on each wing (see page 105).

Common Thrip

Thrips tabaci
Not all thrips have wings, but those that do have tiny wings—like minute feathers—are poor flyers (see page 119).

Green Lacewing

Chrysoperla carnea
You can see from their fine, stringy wings, why lacewings get their name. The bodies of Green Lacewings change colour with the seasons and they are great hunters of aphids and other small flies (see page 104).

BRIMSTONE BUTTERFLY

BROWN, WING EDGE MARKING

Gonepteryx rhamni
LENGTH 6.0–7.4cm (wingspan) · **HABITAT** Grassland, woodland, towns and gardens · **RANGE** Common across the UK and throughout the year
COMPLETE METAMORPHOSIS

THE BRIMSTONE IS A fairly large, pale yellow butterfly, with distinctive, leaf-shaped wings. Adults hibernate through cold weather, so may be seen flying on warm days throughout the year, although they are most common in the spring. Usually seen in ones or twos, they are never very common, but are widespread. They can be found in damp woodlands, along sunny, woodland rides and mature hedgerows, and in large gardens. Brimstone larvae are particularly fond of Buckthorn and Alder Buckthorn.

VIVID ORANGE AND
BROWN PATTERNS

COMMA BUTTERFLY

Polygonia c-album
LENGTH 5–6.4cm (wingspan) · **HABITAT** Open woodland and wood edges · **RANGE** Widespread in England and Wales, rare in southern Scotland and Northern Ireland, from January to December
COMPLETE METAMORPHOSIS

THE COMMA BUTTERFLY IS unmistakable: orange with scallop-shaped wings and brown spots that distinguish it from similar species. Its underside has cryptic brown colouring, making it look like a dead leaf. It gets its name from the comma-shaped white spots on the underside of its wings. The caterpillars feed on Common Nettles, Elms and Willows. They have brown and white flecks that make them look like bird-droppings and help to camouflage them.

SCALLOP-SHAPED
WINGS

DUKE OF BURGUNDY

WHITE FRINGE MARKINGS

Hamearis lucina
LENGTH 1.3–1.8 cm (wingspan) · HABITAT Tall grass
· RANGE Rare, less than 20 sites in southern England
COMPLETE METAMORPHOSIS

THE DUKE OF BURGUNDY

butterfly has dark brown wings patterned with orange spots, with small white dots along the fringes. The females are harder to spot and spend most of their time resting or flying very low to the ground. Living in small colonies, they prefer to reside in areas of taller grass. This scrubby grassland is ideal as the Duke of Burgundy rarely visits flowers. You may see it perching on leaves at the edge of the scrub. It is now a priority species for conservation.

BROWN AND ORANGE PATTERNS

PAINTED LADY

BLACK AND WHITE PATTERN NEAR WING EDGE

Vanessa cardui
LENGTH 4–7.3 cm (wingspan) · HABITAT Parks and gardens
· RANGE Common across Britain from April to October
COMPLETE METAMORPHOSIS

A FAIRLY LARGE ORANGE, black and white butterfly, the painted lady is a migrant to the UK from North Africa, the Middle East and southern Europe during the summer; sometimes it arrives here in enormous numbers. A frequent visitor to gardens, it will feed on Buddleia and other flowers. The caterpillars feed on thistles, mallows and Viper's-bugloss, as well as various cultivated plants. This species cannot survive our winter in any form.

BLACK AND ORANGE PATTERN NEAR THORAX

LARGE WHITE

Pieris brassicae
LENGTH 2.5–3.5 cm (wingspan)
· HABITAT Agricultural areas, meadows and
parkland · RANGE Abundant everywhere
COMPLETE METAMORPHOSIS

MUCH LARGER THAN any of
our other white butterflies,
this species can be
recognised by the black patch
that extends from the wing-tip
to at least halfway along the
outer edge of the wing. The female
has two rounded black spots on both
surfaces of the forewing: the male also
has two spots on the underside, but none on
the upperside. The underside of the hind wing is yellow with a dusting
of black scales. The Large White is a serious pest of cabbages and
sprouts and is often referred to as the Cabbage White.

BLACK SPOTS
ON WINGS

YELLOW TINGE
TO UNDERSIDE

MARBLED WHITE

Melanargia galathea
LENGTH 2–3 cm (wingspan) · HABITAT Rough grassland
· RANGE Found mainly in the southern half of the UK, in
the Midlands and the southwest, from July to August
COMPLETE METAMORPHOSIS

NO OTHER BUTTERFLY found
in the UK or Ireland has this
unmistakable black-chequered
pattern. Eyespots are present
on both surfaces, although not
always obvious on the upperside.
The males are always searching for
newly emerging females. Their
distinctive coloration makes these
butterflies visible even in flight. Adults can
often be seen feeding on purple flowers, such as
Field Scabious, Common Knapweed and Wild Marjoram.

DARK BROWN OR
BLACK AND WHITE
PATTERN

URRY, ORANGE
HEAD

MARSH FRITILLARY

Euphydryas aurinia
LENGTH 3.5–4.6 cm (wingspan) · **HABITAT Open grassland, chalky hillsides, damp meadows and heathland** · **RANGE Found in western Scotland, western England, Wales and many places in Ireland. Thought to be extinct in eastern England, from May to July**
COMPLETE METAMORPHOSIS

THIS BUTTERFLY'S wings have bright, highly variable markings consisting of bands of black and orange with paler orange spots. The wings are darker near the body. The Marsh Fritillary butterfly is found in a range of habitats in which its larval food plant, Devil's-bit Scabious, occurs. Marsh Fritillaries are essentially grassland butterflies in the UK, and although populations may occur occasionally on wet heath, bog margins and woodland clearings, most colonies are found in damp acidic or dry calcareous grasslands.

MORE INTENSE
COLOURS ON
UPPERSIDE OF
WING

MEADOW BROWN

Maniola jurtina
LENGTH 4–6 cm (wingspan) · **HABITAT Parks and gardens** · **RANGE Common from June to September**
COMPLETE METAMORPHOSIS

THE MEADOW BROWN IS ONE of our most common grassland butterflies. It even flies in dull weather when other butterflies are inactive. The Meadow Brown is mainly brown with washed-out orange patches on the forewings. The combination of its relatively large size, orange patches on the forewings only, one eyespot on the forewing and none on the hindwings, is unique to the Meadow Brown. It also has only one small white "pupil" in the eyespots.

COLOURING APPEARS
SLIGHTLY WASHED-OUT

DARK SPOT ON WINGS

ORANGE-TIP

Anthocharis cardamines
LENGTH 4–5.2 cm (wingspan)
· **HABITAT** Hedgerows, woodland rides, meadows, farmland, gardens and parks
· **RANGE** Common, from April to July
COMPLETE METAMORPHOSIS

THE ORANGE-TIP is a common, small butterfly that flies in the spring, between April and July. The male Orange-tip is a white butterfly, half of its forewing is a bold orange, and it has light grey wingtips. The female is also white, but has grey-black wingtips. Both sexes show a "mossy grey" pattern on the underside of their hindwings when at rest. Orange-tip caterpillars are cannibals, eating their own eggshell when they emerge and moving on to eat other Orange-tip eggs nearby.

ORANGE FOREWING
WITH DARKER TIP

UPPERSIDE IS
PLAIN (PATTERN
OF UNDERSIDE
SHOWS THROUGH
AS WING IS SO FINE)

10s
spotters

PEACOCK

JAGGED WING EDGE

Aglais io
LENGTH 6.3–7.5 cm (wingspan) · HABITAT Woodland rides, clearings · RANGE Common, year round
COMPLETE METAMORPHOSIS

THE PEACOCK IS deep red and is so-named for the large blue and yellow "eyes" on each upperwing that bare a marked resemblance to the tail feathers of a peacock. Its underside is dark brown, making the wings look like dead leaves. The peacock is a regular visitor to our gardens where it feeds on buddleia and other flowers. It is on the wing throughout the year, having a single brood, and overwinters as an adult.

MULTI-COLOURED
SPOT ON FOREWING

....

PURPLE EMPEROR

Apatura iris
LENGTH 7–9.2 cm (wingspan) · HABITAT Woodlands or damp ground · RANGE Found in southern England from June to August
COMPLETE METAMORPHOSIS

THE MALE PURPLE EMPEROR has a glossy purple sheen, with white bands across its wings and orange-ringed eyespots under its brown forewings. It is a strikingly beautiful butterfly that can be spotted feeding around the treetops in woodlands. The female is brown but also has orange-ringed eyespots under the forewings. Females are larger than males.

PURPLY-BLUE WING

RED ADMIRAL

Vanessa atalanta
LENGTH 6.4–7.8 cm (wingspan)
● **HABITAT Woods, orchards, hedgerows and parkland** ● **RANGE Common, from January to December**
COMPLETE METAMORPHOSIS

THE RED ADMIRAL is mainly black, with broad, red stripes on the hindwings and forewings, and white spots near the tips of the forewings. Most Red Admirals are migrants to the UK from North Africa and continental Europe, arriving in spring and laying eggs that hatch from July onwards. But some adults manage to survive the winter by hibernating here.

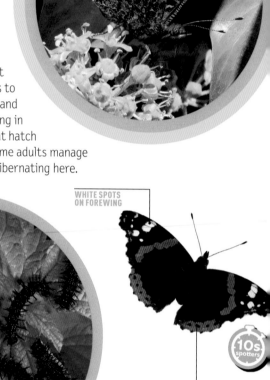

WHITE SPOTS ON FOREWING

RED RIBBON AT EDGE OF HINDWING

10s. spotters

INSECT inspector!

Red Admiral caterpillars eat Stinging Nettles.

DULL BROWN COLOURING ON UPPERSIDE OF WINGS

SMALL BLUE

Cupido minimus
LENGTH 1–1.5 cm (wingspan)
- **HABITAT** Grassland, woodland edges and clearings
- **RANGE** Widely distributed, from May to August
  COMPLETE METAMORPHOSIS

BRITAIN'S SMALLEST

butterfly, the upperside of both the male and female is a sooty brown although the male has a blue sheen. The underside of the wings are silver with dark spots. The male is very territorial and can sometimes be seen chasing other males away. Most colonies are small and consist of just 30 adults. The caterpillars are cannibalistic and will eat up each other if hatched on the same flower, or any smaller larvae that they come across.

10s spotters

FURRY-EDGED WINGS

INSECT inspector!

The Small Blue, like many species of butterfly, has quite different colouring on the upperside and underside of its wings. With the Small Blue the upperside is a dark brown—almost black in the male—with a blue sheen to it, whereas the underside of both males and females is silvery grey. It is thought that the different colorations perform different functions, with the upperside being used mainly for attracting mates while the underside colours are used for camouflage or to warn off predators.

SMALL TORTOISESHELL

Aglais urticae

LENGTH 4.5–6.2 cm (wingspan) ○ HABITAT Grassland, farmland,
woodland, coasts, towns and gardens ○ RANGE It flies almost
everywhere, from October to February
COMPLETE METAMORPHOSIS

THE RICH ORANGE COLOUR and
blue marginal spots readily identify
this very common butterfly. It is a
pretty garden visitor that can be
seen feeding on flowers all year
round during warm spells. Male Small
Tortoiseshells are very territorial,
chasing each other, other butterflies
and anything else that appears in their
space. They court females by "drumming"
their antennae on the females' hindwings.

BEAUTIFUL PATTERN,
WITH BLUE DOTS ALONG
WING EDGES

SPECKLED WOOD

Pararge aegeria

LENGTH 4.6–5.6 cm (wingspan) ○ HABITAT Woodland edges and rides
○ RANGE Found throughout England, Wales and Northern Ireland,
and increasingly in Scotland, from March to October
COMPLETE METAMORPHOSIS

THE APTLY NAMED SPECKLED
Wood flies in partially shaded
woodland with dappled sunlight.
It is dark brown with creamy-
yellow spots. It is the only brown
butterfly with three small,
cream-ringed eyespots on each
hindwing and one on each forewing.
Adults feed on honeydew, while the
caterpillars feed on a variety of grasses,
including False Broom and Cock's-foot. Unlike
any other British butterfly, Speckled Woods can
hibernate as either a caterpillar or chrysalis.

BROWN AND TAN
PATTERNS ON BOTH
SIDES OF WING

SWALLOWTAIL

Papilio machaon

LENGTH 7.6–9.3 cm (wingspan) · HABITAT Wetlands
· RANGE Native to the Norfolk Broads. Migrants may
be spotted in southern England, from May to July
COMPLETE METAMORPHOSIS

THE SWALLOWTAIL IS THE

UK's largest butterfly. It is striking and exotic-looking —its creamy-yellow wings have heavy black veins and blue margins. Its hindwings have distinctively long "tails" and a red spot. Adults fly between May and July when they can often be seen over reedbeds, or feeding on Ragged-robin or flowering thistles. The caterpillars feed on Milk-parsley.

IRREGULAR SHAPED YELLOW
SECTIONS WITH BROWN EDGES

POINTED WING TIP

INSECT inspector!

When threatened, Swallowtail caterpillars inflate a fleshy, orange organ, called an osmeterium, from behind their heads. It looks like a snake's forked tongue and gives off a pineapple-like smell.

WHITE ADMIRAL

Limenitis Camilla

LENGTH 5.6–6.6 cm (wingspan) • HABITAT Woodland
• RANGE Widespread in southern England, from June to August
COMPLETE METAMORPHOSIS

THE WHITE ADMIRAL is a black butterfly with distinctive white bands on the wings and a gingery-brown underside. It has a characteristic flight pattern of short periods of wingbeats followed by long glides. Adults are often found on the flowers of Bramble and lay their eggs on Honeysuckle leaves, which the caterpillars feed on. Usually seen in ones or twos, it is never very common, but is widespread in southern England.

FURRY, GREY THORAX

LIGHTER BROWN COLOURING ON WING UNDERSIDE

INSECT inspector!

It is a common misconception that moths only fly at night and butterflies fly by day, as there are many day flying moths. It is also true that many moths are as colourful as many butterflies, such as the Scarce Forester shown here. It is a wonderful turquoise blue colour and is often seen flying in the daytime. It is true to say that in the UK and Ireland, moths are generally fluffier and have fluffy antenna whilst butterflies are smooth, but even this difference breaks down in other parts of the world.

CINNABAR

Tyria jacobaeae
LENGTH 3.4–4.6 cm (wingspan) · HABITAT Rough grassland and hedgerows, gardens and waste ground · RANGE Widespread, from May to August
COMPLETE METAMORPHOSIS

THE CINNABAR IS A dark brown moth with two red spots and two pinky-red stripes on its rounded forewings. Its hindwings are pinky-red and bordered with black. Its black-and-yellow-banded caterpillars feed on Common Ragwort. They pupate in autumn, spending the winter as cocoons on the ground, before emerging as moths the following summer. The bright colouring of both caterpillar and adult warns predators that they are unpalatable, having ingested the poisonous Ragwort plants.

VIVID, RED STRIPE

ELEPHANT HAWK-MOTH

Deilephila elpenor
LENGTH 4.5–6 cm (wingspan) · HABITAT Parks, gardens, woodland edges, rough grassland and sand dunes · RANGE Common, from May to August
COMPLETE METAMORPHOSIS

THE ELEPHANT HAWK-MOTH is a medium-sized hawk-moth and is so-called because the caterpillar looks a little like an elephant's trunk. The caterpillars are greyish-green or brown, with two enormous, black eyespots towards the head. As protection from predators, they swell up to show these spots. The caterpillars feed on Willowherbs, Fuchsia and Bedstraw, and the adults feed on nectar. Adults are mainly golden-olive with bright pink bars on the wings and body. They are active at dusk.

PINK STRIPES ON WINGS

LARGE THORAX AND ABDOMEN

HUMMINGBIRD HAWK-MOTH

ORANGE HINDWING

Macroglossum stellatarum

LENGTH 5–5.8 cm (wingspan) · HABITAT Woodland edges, heathland and scrub
· RANGE A widespread migrant, most frequent in the south and near the
coast, from April to December
COMPLETE METAMORPHOSIS

THIS SMALL, DAY-FLYING moth is usually seen as a brownish blur, hovering like a Hummingbird as it feeds with its long proboscis on the nectar of Honeysuckle, Red Valerian and other flowers. It has greyish-brown forewings, bright orange hindwings, and a greyish body with a broad, black-and-white "tail". Its hovering flight is a distinguishing feature; it flutters its wings so quickly that it makes an audible hum.

LONG PROBOSCIS

THICKER LEG TOP

LUNAR HORNET CLEARWING

Sesia bembeciformis

LENGTH 3.2–4.2 cm (wingspan)
· HABITAT Woodland, commons and damp meadows
· RANGE Found throughout the UK, from June to August
COMPLETE METAMORPHOSIS

LUNAR HORNET CLEARWING

moths get their name from their hornet-like colouring. To add to the effect, they also have clear wings. After emerging from their cocoons, the young moths lose most of their wing scales leaving a transparent window on each wing. The caterpillars feed mainly on the roots of Sallow, Poplar and Willow trees. Eggs are laid on the bark, and the larvae burrow into the wood.

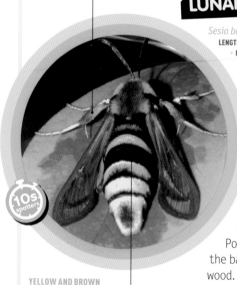

YELLOW AND BROWN HORNET-LIKE COLOURING

PUSS MOTH

Cerura vinula
LENGTH 5–8 cm (wingspan) • **HABITAT** Woodland, park areas
or even in your back garden. • **RANGE** Widespread
throughout the UK, from May to July
COMPLETE METAMORPHOSIS

THE PUSS MOTH is covered with soft cat-like fur, hence its name. Its striking black and grey marbled markings make it easy to identify. When the caterpillar first emerges, it's black with what looks like a long double tail extending from its abdomen. The caterpillar goes through several changes known as "instars". Within a short period, the caterpillar turns into a larger, green caterpillar with a dark coloured patch on its back and a red ringed cartoon-like face. Before it is ready to pupate, it changes to orange and then purple, spins a cocoon of silk around itself and uses bits of bark as camouflage to keep hidden.

INSECT inspector!

If agitated, the Puss Moth caterpillar will wave around two long whip-like appendages from its tail. If further deterrent is needed, the caterpillar will squirt formic acid, like that found in bee stings, from its thorax.

CAT-LIKE FUR

DARK WAVE PATTERN ON WING

10s spotters

SIX-SPOT BURNET

Zygaena filipendulae

LENGTH 3–3.8 cm (wingspan) · **HABITAT** Grassland, woodland rides and sand dunes · **RANGE** Widespread in England and Wales, rarer in Scotland where it is mainly found near the coast, from June to August
COMPLETE METAMORPHOSIS

THE SIX-SPOT BURNET is a day-flying moth. Its forewings are grey-black with a strong blue or green sheen, and each has six red spots. The spots indicate to predators that they are poisonous: they release hydrogen cyanide when attacked. Adults feed on the nectar of Knapweed and Thistles, and females lay their eggs on the caterpillars' food plants. The caterpillars hatch and feed, hibernating over at least one winter. They emerge the following spring and pupate in a papery cocoon attached to grass stems.

GREY-BLACK WING WITH BLUE OR GREEN SHEEN

6 RED SPOTS ON EACH FOREWING

VAPOURER

Orgyia antiqua

LENGTH 1.2–1.7 cm (wingspan) · **HABITAT** Woodland, heathland, moorland, hedgerows, parks and gardens · **RANGE** Widespread and common across much of the United Kingdom, although scarce in some western areas, from July to October
COMPLETE METAMORPHOSIS

ADULT MALE VAPOURERS have broad orange-brown wings, with a few faint lines crossing them, an obvious white eyespot near the corner of each forewing and large, comb-like antennae. Females are almost wingless, plump, buff-coloured and with smaller, non-combed antennae. The hairy caterpillars are grey, black and red, with tufts of yellow hairs running along their back and two black tufts at the front that almost give them the impression of a hairy scorpion.

COMBED ANTENNAE

WINTER MOTH

Operophtera brumata

LENGTH 2.2–2.8 cm (wingspan)
- **HABITAT** Woodland, hedgerows, orchards, gardens
- **RANGE** Widespread, from October to February

COMPLETE METAMORPHOSIS

MALE WINTER MOTHS have dark cross bands on their wings whereas females are totally flightless with tiny dark-striped wings allowing them to be easily distinguished from each other. Females generally spend the day at the base of trees, climbing up at dusk to attract a mate and lay eggs. The female can be carried by the male if disturbed during mating. Females lay eggs on a wide range of broad-leaved trees and shrubs. The white-striped green caterpillars hatch in spring and "balloon" from tree to tree by spinning a silk thread and using the breeze to carry them along to the next food source.

MALE HAS LIGHT BROWN WINGS

FEMALE HAS TINY WINGS

INSECT Inspector!

Blue Tits time their breeding to coincide with the emergence of Winter Moth caterpillars in spring. A single brood of Blue Tits can eat up to 10,000 caterpillars!

INSECT REPORT: INVASIVE SPECIES

Argentine Ant

Linepithema humile

These ants started out in South America but have now spread over the whole world! Because they are from a single mega-colony they don't fight each other but they do attack native UK ants.

CATERPILLARS OF THE OAK PROCESSIONARY MOTH
(THAUMETOPOEA PROCESSIONEA) NEST HIGH UP ON AN OAK
TRUNK. THEY ARE AN INVASIVE NON-NATIVE SPECIES FROM
SOUTHERN EUROPE UNWITTINGLY INTRODUCED TO ENGLAND AT
THE TURN OF THIS CENTURY. THEY GET THEIR NAME FROM THE
WAY THE CATERPILLARS CRAWL IN LONG LINES—PROCESSIONS.
WHEN THEY NEST IN GREAT NUMBERS, THEY CAUSE DEFOLIATION
OF THE OAKS AND THEIR HAIRS CAN CAUSE A RASH IN HUMANS.

People move from place to place—so do insects. This isn't always good. Sometimes insects cause harm when they are accidentally introduced to new environments. They can crowd out native species, kill native plants and animals, or even pose a threat to people living in the area.

Harlequin Ladybird

Harmonia axyridis

Originally from Asia, this aggressive predator can be a gardener's friend as it eats aphids—unfortunately it also eats many other insects too, including eggs and larvae of butterflies, moths and even other ladybirds. It can even bite humans! Since its introduction into the UK in 2004 it has spread very swiftly across much of the country. Adults often invade people's homes for the winter. Harlequin Ladybirds are considered one of the most invasive animal species worldwide (see page 18).

sian Hornet

Vespa velutina

Originally from China, this insect has invaded most of Europe after accidentally being introduced into France about 15 years ago. It has only inhabited the UK since 2016, but it is a big pest of Honey Bees. It sits outside hives and waits to ambush worker bees. This insect has more of a liking of Honey Bees than the UK's native hornet (see page 76).

SCORPION FLY

Panorpa communis
LENGTH Up to 30 mm • **HABITAT** Woods, gardens and hedgerows
• **RANGE** Widespread from May to September
COMPLETE METAMORPHOSIS

THE SCORPION FLY is so-called because of its scorpion-like tail, which does not sting, but is used by the male fly in courtship displays. The adults usually mate at night but during mating, the male is at risk of being killed by the female! The female lays her eggs in the soil. Scorpion Flies are usually found in gardens, woodlands and hedgerows, especially in Brambles and Stinging Nettles. They have yellow and black bodies, with a reddish-coloured head that is shaped like a long, pointy beak, which they use for scavenging food. They often take dead insects from spiders' webs to feed on.

10s spotters

TAIL RESEMBLES SCORPION STINGER

SNOW FLEA

LONG, BEAK-SHAPED HEAD

Boreus hyemalis **LENGTH** 5 mm • **HABITAT** Lives in moss
• **RANGE** Widespread, seen in the winter months
COMPLETE METAMORPHOSIS

THE SNOW FLEA is not a flea but it can jump! It lives among moss but it is most easily seen when it walks over snow—hence its name. Its head has a very long downward projecting "beak"—like the scorpion fly—and its body is dumpy, with dark metallic reflections. The Snow Flea is adult in the winter and rather than flying, it crawls and jumps—the male wings are modified into curved spines used in mating.

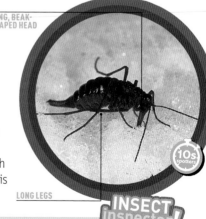

10s spotters

LONG LEGS

INSECT inspector!

It is not known how Snow Fleas manage to jump so high as they do not have muscular hind legs—however, they can jump up to 5 cm!

ANTLION

Euroleon nostras
LENGTH 30 mm, wingspan 70 mm, larvae 10 mm
· HABITAT Sandy dunes, hollow trees, forest floors, under hedges, and in dark, shady areas
· RANGE Norfolk and Suffolk Sandlings
COMPLETE METAMORPHOSIS

ANTLIONS HAVE SOME pretty interesting nicknames—everything from "demons in the dust" to "doodlebugs." And it's all because of the show they put on while trying to catch prey. Antlions dig pits in the ground to trap ants. Only the best places will do, so these insects meander along searching for the perfect spot to dig. They leave a path that looks like an aimless doodle in their wake. The larvae require dry sandy soil to dig their pits, close to vertical sandy ledges that help adults emerge. Larval Antlions have such efficient digestion that they do not produce solid waste and therefore do not need an anus. Larvae exude only liquid waste; the small amount of solid waste that may build up is excreted by newly emerged adults. Antlions remain in their larval stage for two years before pupating. Adults emerge from the pupa towards the end of July or in the first few days of August. They gather in a tall pine tree, and a number of males attempt to attract a single female. After mating, the female flies to the ground, where she lays her eggs in the sand. She has to be particularly wary of Antlion larvae at this time, which are the main predators of female adults. Males live for up to 20 days, while females last a little longer, with an average life span of 24 days.

LONG, NARROW ABDOMEN

FINELY VEINED WINGS

BORDERED BROWN LACEWING

CLEAR WINGS, WITH BROWN SPECKLES

Megalomus hirtus
LENGTH 6.5–8.5 mm (wingspan) · **HABITAT** Two sites in Scotland · **RANGE** Restricted, from June to August
INCOMPLETE METAMORPHOSIS

THE BORDERED BROWN LACEWING is currently only known in two sites in Scotland, Holyrood Park in Edinburgh and Muchalls in Aberdeenshire. As a result of its rarity, the lacewing is classified as a UK Biodiversity Action Plan (UKBAP) priority species and in Scotland is on the Scottish Biodiversity List. The Bordered Brown Lacewing feeds on Aphids and adults have four, intricately patterned wings.

GIANT LACEWING

TRANSPARENT, FINELY VEINED WINGS

Osmylus fulvicephalus
LENGTH 25 mm · **HABITAT** Streamside vegetation, especially in woods · **RANGE** South and west England and Wales, April to August
INCOMPLETE METAMORPHOSIS

THE LARGEST OF THE lacewings found in the UK and Ireland, Giant Lacewings are typically light brown with black spots on their wings. Their larvae are amphibious and eat other small, water-dwelling creatures.

GREEN LACEWING

Chrysoperla carnea **LENGTH** 10 mm · **HABITAT** Gardens, fields, hedges, woodland, trees and bushes · **RANGE** All year round but particularly May to September
INCOMPLETE METAMORPHOSIS

WITH THEIR LONG ANTENNAE, golden eyes and two pairs of transparent wings that are nearly twice as long as their abdomen, these dazzling creatures can be found pretty much everywhere in high volume. Despite their name, green lacewings do vary in colour and some are striking shades of blue. They hibernate in winter and change colour to a yellowish-brown in the warmer months.

LONG WINGS

BANDED DEMOISELLE

SHINY, BLUE THORAX

DARK "SMUDGE" ON WINGS

Calopteryx splendens · LENGTH 4.5 cm · HABITAT Slow-flowing streams and rivers · RANGE Common
INCOMPLETE METAMORPHOSIS

THE MATURE MALE Banded demoiselle has a brilliant metallic blue body and a dark blue patch on the outer part of each wing. The female is metallic green, usually becoming bronze with age, and has yellowish-green wings. The insect prefers slow-moving rivers and canals with muddy bottoms. Females can lay up to ten eggs per minute for 45 minutes. The eggs hatch after 14 days. The larvae have very long legs and are stick-shaped.

INSECT inspector!

Damselflies and dragonflies have been around for 400 million years.

COMMON BLUE DAMSELFLY

TRANSPARENT WINGS

Enallagma cyathigerum
LENGTH 3.2 cm · HABITAT Any waterbody, grassland and woodland · RANGE Widespread, from April to September
INCOMPLETE METAMORPHOSIS

THE MALE COMMON BLUE Damselfly is pale blue with bands of black along its body; the female is either blue or dull green, with distinctive black "torpedo" markings. The Common Blue Damselfly is our most common damselfly. It is an aggressive species—males will defend their females as they lay their eggs, both from their own kind and other species. When Common Blues mate, they form a "mating wheel" in which the male clasps the female by the neck and she bends her body around to his reproductive organs.

SEGMENTED ABDOMEN

LARGE RED DAMSELFLY

Pyrrhosoma nymphula

LENGTH 3.5–4.5 cm · **HABITAT** Freshwater habitats of all kinds, including peat bogs and muddy ditches; sometimes breeds in brackish water · **RANGE** Widespread, from April to September
INCOMPLETE METAMORPHOSIS

THIS SPECIES HAS A bright red abdomen with black marks in both sexes, but the male tends to be more vibrant than the female. The latter has a thin black line down the middle of the abdomen. Both sexes have red stripes on a black thorax, although the stripes are yellowish at first. Look also for the black legs. The male is aggressive, fighting off other males entering his territory. Damselflies rest with their wings folded lengthways along their body.

GOLDEN THORAX STRIPE

BLACK LEGS

COMMON DARTER

Sympetrum striolatum

LENGTH 3–4 cm · **HABITAT** Almost anywhere, but most common by water · **RANGE** Abundant in England, Wales and Ireland but less common in Scotland, from May to October
INCOMPLETE METAMORPHOSIS

THE MALE HAS A BRICK-RED abdomen, somewhat duller than that of other red darters. Look for the tiny black dots, surrounded by yellow, towards the rear of each abdominal segment. The female is yellowish-brown. There are clear yellow stripes on the legs of both sexes. This species breeds in either still or slowly moving, shallow water.

COLOURLESS WINGS

YELLOW STRIPES ON LEGS

EMPEROR DRAGONFLY

DARK SPOT ON EACH WING

Anax imperator

LENGTH 7.8 cm ∘ HABITAT Large ponds and lakes, as well as canals and ditches ∘ RANGE Found in Central and Southern England and South Wales, from June to August
INCOMPLETE METAMORPHOSIS

THE EMPEROR DRAGONFLY is a very large, impressive dragonfly. Males are pale blue, with an apple-green thorax and a black stripe running the length of the body. Females are similar, but a duller greeny-blue. Both have greeny-blue eyes. This species is recognised by the combination of its large size and mostly blue colour. The Emperor Dragonfly flies up high to look for insect-prey, such as butterflies and chaser dragonflies. It catches its prey in mid-air and may eat it on the wing.

10s spotters

BLACK AND BLUE PATTERNING ALONG LENGTH OF ABDOMEN

FOUR-SPOTTED DARTER

Libellula quadrimaculata

LENGTH 4 cm ∘ HABITAT Large ponds, slow-flowing rivers and streams ∘ RANGE Widespread across Britain and Ireland, more numerous in southern areas
INCOMPLETE METAMORPHOSIS

ALSO KNOWN AS the Four-spotted Chaser, this dragonfly is found all over the northern hemisphere. It lives near ponds and slow-flowing streams and its larvae feed on other aquatic insects. The adult eats small flies. Its main predators are birds and other larger dragonflies, like the Emperor.

YELLOW SIDES TO ABDOMEN SEGMENTS

BLACK TAIL END OF ABDOMEN

GOLDEN-RINGED DRAGONFLY

Cordulegaster boltonii
LENGTH 7.4–8.4 cm · HABITAT Small acidic streams in moorland and heathland · RANGE Found in west Scotland, Cumbria, Southern England and Wales, from May to September
INCOMPLETE METAMORPHOSIS

THE FEMALE OF THIS species is the UK's longest dragonfly because of her long ovipositor. Golden-ringed Dragonflies are voracious predators, feeding on damselflies, other dragonflies, wasps, beetles and bumblebees. They are fast, agile and powerful flyers. Both sexes are black, with yellow bands along the body and bright-green eyes. Males have a shorter body and are stouter than the females.

WASP-LIKE COLOURING

10s spotters

BLACK LEGS

INSECT inspector!

The larvae live buried at the bottom of streams, ambushing prey as it passes by. They grow very slowly and may spend up to five years in the water before they emerge onto the land, shedding their nymph body and emerging as a dragonfly.

MIGRANT HAWKER

Aeshna mixta

LENGTH 6.3cm · **HABITAT** Gardens, grassland and woodland
· **RANGE** Found in Southern and Central England and South Wales, from July to November
INCOMPLETE METAMORPHOSIS

THE MIGRANT HAWKER is mostly dark brown and black in colour. The male has pale blue spots and yellow flecks all along the body, dark blue eyes, and pale yellow-and-blue patches on the thorax. The female has yellowish spots and brownish eyes. The Migrant Hawker is smaller and has more brown on it than the other three large species (Common, Azure and Southern Hawkers). Hawkers are the largest and fastest flying dragonflies; they catch their insect-prey mid-air and can hover or fly backwards.

BROWN COLOURING MIXED WITH BLUE AND BLACK AT TOP OF ABDOMEN

RED THORAX

WHITE-FACED DARTER

Leucorrhinia dubia

LENGTH 3.3–3.7 cm · **HABITAT** Lowland peatbogs
· **RANGE** Isolated sites in the English Midlands and the Scottish Highlands
INCOMPLETE METAMORPHOSIS

ONE OF THE RAREST dragonflies in the UK, the White-faced Darter can only be found in shallow peaty pools in the lowland peat bogs of the Scottish Highlands and the English Midlands. Territorial, it can often be seen roosting on trees and bushes up to 50 metres away from a bog. It can be spotted during the summer months catching smaller prey on the wing, consuming them on a perch away from the bog.

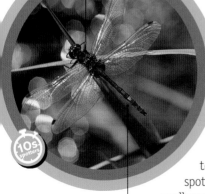

2 OR 3 ORANGE SECTIONS ON OTHERWISE BLACK ABDOMEN

BOG BRUSH CRICKET

Metrioptera brachyptera
LENGTH 1.2–1.6 cm ▪ **HABITAT Bogs and wet heath**
▪ **RANGE From July to October**
INCOMPLETE METAMORPHOSIS

THE BOG BRUSH CRICKET is brown or green above and bright green below. The wings are normally short in both sexes. The song, heard mainly by day, is a series of shrill clicks, like the rapid ticking of a clock. The species feeds on herbs and some small insects.

BROWN UPPERSIDE

LONG LEGS

GREAT GREEN BRUSH-CRICKET

ORANGEY-BROWN STRIPE

Tettigonia viridissima
LENGTH 7 cm ▪ **HABITAT Grassland, woodland and moorland**
▪ **RANGE Found in Southern England and South Wales, from May to October**
INCOMPLETE METAMORPHOSIS

THE GREAT GREEN BUSH-CRICKET is by far our largest bush-cricket. It is green with an orangey-brown stripe running the length of the body, and long wings. It lives in trees and on grassland dotted with patches of scrub. It prefers light, dry soils where females can lay their eggs using their very long, down-curved ovipositors. Males display to females by rubbing their forewings together to produce a very loud, long "song"; but their expert camouflage still makes them hard spot.

ORANGE ANTENNAE

HOUSE CRICKET

Acheta domesticus
LENGTH 1.4–2 cm · HABITAT Heated buildings, especially in and around kitchens, and rubbish dumps: may move into the surrounding countryside in hot summers · RANGE Widespread in England and Wales, scarcer in Scotland, from January to December
INCOMPLETE METAMORPHOSIS

BOTH SEXES OF this greyish or brown cricket are fully winged and can fly. Originally from North Africa, the insect is found mainly in permanently-heated buildings, such as bakeries, where there is plenty of food, but it can also be found out of doors in hot summers and it can survive throughout the year on large rubbish dumps where fermentation provides the necessary warmth. Its shrill, whistle-like song is heard mainly at night.

THICK UPPER LEGS

PRONOUNCED "COLLAR"
AT TOP OF THORAX

INSECT Inspector!

Crickets are often used as food for pet reptiles, but many people think humans need to be looking at insects like crickets as potential future food for humans too. Many insects are abundant and are high in protein.

OAK BUSH-CRICKET

Meconema thalassinum
LENGTH 1.3–1.7 cm • **HABITAT** Ancient woodlands, hedgerows, parkland and gardens • **RANGE** Found in England and Wales; common in the south and Midlands, but absent in the north, from June to December
INCOMPLETE METAMORPHOSIS

THE OAK BUSH-CRICKET has a slender, lime-green body, with medium-length wings and an orangey-brown strip running down its back. It can be found in the canopy of mature trees. The Oak Bush-cricket is not often seen, but does fly well and is attracted to lights at night. It does not have a "song" as such, but drums on leaves with its hind legs. The female lays her eggs in tree bark in late summer and the nymphs emerge the following June. Female Bush-crickets can be distinguished from their male counterparts by their long, curved ovipositors, visible at the end of the body. Males have two short, rounded claspers.

ORANGE EY

WART-BITER BUSH-CRICKET

Decticus verrucivorus
LENGTH 3.1–3.7 cm • **HABITAT** Rough grassland, heaths and marshes • **RANGE** Confined to a few sites in Southern England, from July to October
INCOMPLETE METAMORPHOSIS

THE WART-BITER HAS BIG, dark eyes and is typically dark green, often with dark brown or black blotches on the body and wings. Its large and powerful hind legs give this species a frog-like appearance. It "sings" or stridulates by rubbing its wings together. Its loud and distinctive song consists of a series of rapidly repeated clicks in short bursts and often lasts for several minutes—it sounds a bit like a free-wheeling bicycle. Wart-biters rarely fly as they are too heavy and their wings are not large enough; most can only fly very short distances.

BROWN BLOTCHES

SPINY LEGS

COMMON FIELD GRASSHOPPER

Chorthippus brunneus
LENGTH 1.8–2.4 cm
° **HABITAT Open, sunny, grassy areas, including gardens**
° **RANGE Widespread, from May to October**
INCOMPLETE METAMORPHOSIS

THE ADULT COMMON FIELD
Grasshopper is usually mottled brown in colour, with barring on the sides, but various parts of the body and wings may be grey, purple, black or green. Nymphs can be pink in colour. It is most easily identified when seen up close as its very hairy underside becomes visible. The tip of the male abdomen may be red or orange. Males can be seen displaying to females by rubbing their legs against their wings to create a "song"—in this case, it is a brief, single chirrup, repeated at short intervals, which resembles time-signal pips.

10s. spotters

USUALLY BROWN BODY, BUT CAN COME IN GREEN OR PURPLE COLOURS

LARGE, HAIRY HIND LEGS

USUALLY GREEN
BODY COLOUR

PRICKLY STICK INSECT

Acanthoxyla geisovii

LENGTH 8–10 cm • **HABITAT** Hedgerows, gardens and roadsides
• **RANGE** Rare, parts of Devon, Cornwall and the Isles of Scilly only
INCOMPLETE METAMORPHOSIS

THERE ARE NO NATIVE stick insects in the UK but several species have been introduced over the years. The Prickly Stick Insect is one such species that is thought to have come here in ferns from New Zealand about 100 years ago—it is still only found in the south of England. Prickly Stick Insects are covered with black spines. Females produce eggs—just dropping them at their feet—without any interaction from a male, in fact no male Prickly Stick Insects have ever been found!

BLACK SPINES

UNARMED STICK INSECT

Acanthoxyla inermis

LENGTH 10 cm • **HABITAT** Hedgerows, gardens and roadsides
• **RANGE** Rare, southern England
INCOMPLETE METAMORPHOSIS

THE UNARMED STICK INSECT has a long thin body and three pairs of thin, jointed limbs. Its long thin body is very hard to make out from real sticks, which allows these insects to hide in plain sight. As well as appearing like a small tree branch, stick insects also behave like a part of the tree, performing swaying motions to give the appearance of moving in the breeze. It is another species that is native to New Zealand.

SMOOTH BUT
SEGMENTED BODY

THIN LEGS

NORTHERN FEBRUARY RED

Brachyptera putata
LENGTH 20 mm · **HABITAT Rivers that are in open heaths or upland pastures** · **RANGE Mainly northeast Scotland and the Highlands, from February to April**
INCOMPLETE METAMORPHOSIS

THE NORTHERN FEBRUARY RED is a stonefly that likes rivers with good water quality and lots of winter sunshine, especially those that run through open heaths or upland pastures. Adult males have short wings, making them poor at flying. Adult females are much better at flying and can move back upstream to find a mate. Nymphs do not fly and are usually found among stones in rivers. Winter sunlight encourages the growth of different types of algae which is food for the larvae. Cold water also produces high oxygen levels which helps the larvae and nymphs to remain active during the winter months.

MOTTLED WINGS

The Northern February red is one of very few insects that are only found in the UK. This means it is one of what are known as endemic species. Sadly this species is currently in decline. Many of the other insects that are endemic to the British Isles are located on small islands, like the Scilly Bee—found in the Isles of Scilly, off the Southwest tip of Cornwall, and the Lundy Cabbage Flea Beetle—found on Lundy, an island in the Bristol Channel.

SNAKEFLY

Raphidiidae spp.
LENGTH 15 mm ○ **HABITAT Coniferous forests**
○ **RANGE Widespread and fairly common**
COMPLETE METAMORPHOSIS

THE ADULT SNAKEFLY
resembles a lacewing
with a protruding
"snake-like" head. It is a
predatory insect that
lives in trees and feeds
on aphids and other
small flies. Snakefly
larvae eat the eggs and
young of other insects
and arthropods. Adult
Snakeflies themselves are
eaten by many different
predators, especially woodland birds
like Woodpeckers and Treecreepers, while
their young are often taken by wasps. Snakeflies go
through one of the more numerous stages of development in the insect world, as the larva can moult 10 or
more times before finally creating a pupa from which the
adult emerges.

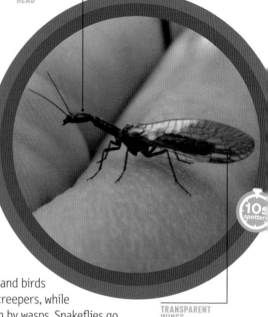

SNAKE-SHAPED
HEAD

TRANSPARENT
WINGS

INSECT inspector!

Because they spend almost all of their life in the
tree canopy, sightings of Snakeflies are quite
rare—even though they are not endangered in the
UK. Fossils have shown that there used to be many
more species of Snakefly around the world but
many of them died out during the same historical
event which wiped out the dinosaurs.

ALDER FLY

Sialis lutaria
LENGTH 14 mm · **HABITAT Vegetation near to water**
· **RANGE Widespread, from January to December**
COMPLETE METAMORPHOSIS

ADULT ALDER FLIES are blackish-brown, with dark, lacy wings which they fold in a tent-like manner along the length of their body. The larvae are aquatic carnivores that resemble slugs and live in the silt at the bottom of ponds and slow-flowing rivers. The larvae take around 24 months to develop before emerging onto land for their pupal stage. Adults emerge in large numbers during the summer and live for only a few weeks during which they are focused on reproducing. Mating happens at night and the females lay their eggs on branches that overhang water bodies, so that once the larvae emerge they can simply drop into the water and start their underwater life.

LONG ANTENNAE

SHORT LEGS

Laugh Out Loud!

Joke: Did you hear about the flies playing football in a saucer?

Answer: Next week, they are hoping to be playing in the Cup!

Simuliidae spp.

LENGTH 3–7 mm • **HABITAT Anywhere there is running water**
• **RANGE Common and widespread across UK and Ireland**
COMPLETE METAMORPHOSIS

ALTHOUGH THEY ARE SMALL

these insects represent one of the larger problem insects for humans due to the fact that they bite us and can spread disease. Blackflies from the *Simulidae* family—not to be confused with the Blackflies or Black Aphids *(Hemiptera: Aphididae)* on page 52, which share their common name—are found wherever there is water. The adult fly lays its eggs in or near the water and then the larva hatches in the form of a long segmented body. They are particularly abundant where the water flows quickly. Blackfly are only in their larva form for a week or so before they pupate and then emerge as adults. The adult lives for a similar amount of time so that the whole life cycle of the Blackfly might only last from 3 weeks to a month.

INSECT inspector!

The female Blackfly requires blood as part of the development of her eggs. For this reason they will seek out warm-blooded animals—including humans—to bite. When they bite they inject saliva into the skin of the host which is why we sometimes get swelling and irritation when we are bitten. While this irritation can be sore and annoying, in some countries it can be much worse as the Blackfly can spread serious diseases like river blindness which as the name suggests can lead to the victim losing their sight.

ROUNDED, BULBOUS THORAX

10s spotters

TRANSPARENT WINGS

COMMON THRIP

Thrips tabaci
LENGTH up to 1 mm · **HABITAT Grassland and farmland** · **RANGE Common and usually seen on humid summer days**
INCOMPLETE METAMORPHOSIS

THRIPS ARE TINY WINGED or wingless black or brown insects. The wings, when present, are small and feathery, but even where they are present Thrips are poor flyers. Most thrips are vegetarians, sucking sap from plant cells with their piercing beak. They have asymmetrical mouthparts with the right side of their mouth being almost redundant during feeding while the left side is used to cut into plants and extract the nutrients. Thrips are considered a pest as they can damage crops, with Common Thrips being particularly partial to onions, tomatoes and cabbages. As well as eating these plants, Thrips often carry viruses— like the Tomato Spotted Wilt Virus— that also damage the plants.

LONG, THIN
BODY SHAPE

INSECT inspector!

Thrips have a variety of common names that refer to their behaviors. These include Picture Frame Flies —they are often found in large numbers between glass and picture, and Thunderflies because they tend to appear in warm humid weather when thunder is imminent.

GREAT RED SEDGE

Phrygania grandis
LENGTH 18–28 mm
- **HABITAT** Around still and slow-moving water
- **RANGE** Widespread and frequent in much of England, but fewer records from Scotland and Wales, from May to August
COMPLETE METAMORPHOSIS

THE FEMALE GREAT RED

Sedge is the largest British caddisfly and has a dark stripe on the forewing which is not present in the male. The male is also the smaller of the sexes. These are flies of lakes and slow-flowing rivers, and they hatch on summer afternoons and evenings. The larva makes its case from plant material, which it fashions into a spiral tube. After pupating, the adult emerges and heads for land, where it rests among the stones and low bushes.

LONG ANTENNAE

MOTTLED BROWN PATTERN COVERS HEAD, THORAX, WINGS AND LEGS

INSECT inspector!

There are over 200 types of caddisfly in the UK. The larvae of almost all of these species (other than the Land Caddisfly), live underwater. To protect themselves from predators, many caddisfly larvae build shelters called cases that they make by combining underwater debris—like sand, rocks, seeds and twigs—with a silk they make themselves. Each species of caddisfly makes its case in a specific way and using certain debris types, so that you can tell which species made an individual case. Some of these cases are fixed in place but others can be moved by the larva, almost like a Hermit Crab with its shell.

SILVERFISH

Lepisma saccharina
LENGTH 13–25 mm • **HABITAT Houses**
• **RANGE Widespread**
INCOMPLETE METAMORPHOSIS

A FAST-RUNNING, wingless insect, covered with silvery scales, that lives mainly in houses, especially in dark cupboards and on undisturbed bookshelves. It prefers slightly damp spots and is mainly nocturnal. The Silverfish feeds on moulds and starchy materials, including book-bindings and the glue of cartons, and can cause serious damage to books and other papers. Its common name derives from the animal's silvery light-grey colour, combined with the fish-like appearance of its movements. They are often considered a pest in houses because an adult Silverfish can lay up to 50 eggs in one go, and once they hatch, Silverfish can live for up to 8 years, so if you are unlucky enough to find a few Silverfish in your carpet, you know there are probably a lot more very close by.

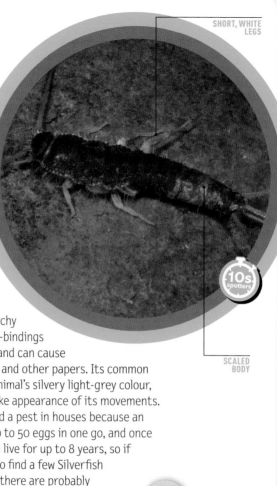

SHORT, WHITE LEGS

SCALED BODY

10s spotters

Laugh Out Loud!

Joke: Did you hear about the light that turned into an insect?

Answer: It was a larva lamp!

INSECT REPORT: CONSERVING BENEFICIAL INSECTS

Say "DON'T SPRAY"

Insects are key players in helping your garden and our ecosystem thrive! You could ask grown-ups to be mindful when using pesticides and chemicals in the garden. The same chemicals that kill pests can also kill beneficial insects.

Don't be rubbish!

Many insects live in or around water. Anything you drop usually ends up in a river or stream, so make sure you never drop litter. If you have some spare time, you might join a litter pick where groups get together in public places to pick up rubbish.

Human activity is constantly changing the world we live in. Some of the things we see as development can have a harmful effect on the environment, but there are small things we can all do to make it easier for insects to thrive.

MANY INSECTS INHABIT DEAD WOOD AND ROTTING TREE STUMPS. ENCOURAGE ADULTS IN YOUR COMMUNITY TO LEAVE OLD TREES AND SHRUBS TO BREAK DOWN NATURALLY. IF YOU GET A CHANCE, MAKE YOUR OWN INSECT HABITATS LIKE THIS OLD BIRD HOUSE FILLED WITH HOLLOW CANES AND FIR CONES

Glossary

ABDOMEN: The part at the end of an insect's body. It contains the respiratory, digestive and reproductive systems.

ADAPTATION: A specialised body part or behaviour that allows an animal to survive and reproduce in its environment

ANTENNA: A long, thin organ that protrudes from an insect's head. Insects use them to feel, smell and, in some species, to hear.

ARTHROPOD: An invertebrate animal with a segmented body, jointed appendages, and an exoskeleton. Insects are arthropods

CAMOUFLAGE: A natural disguise, such as skin colour or pattern, that helps an animal blend in with its surroundings

COLONY: A group of the same kind of organism living or growing together

EGG CASE: A capsule that contains eggs

ELYTRA: A pair of hardened forewings found on some insects, especially beetles. They are not used for flying, but to protect the delicate hind wings.

ENDANGERED: A plant or animal that is at risk of becoming extinct, or no longer existing

EXOSKELETON: An external supportive covering on an animal's body

FOREWINGS: The two front wings of a four-winged insect

GRUB: The short, fat, worm-like larva of certain insects, such as beetles and wasps.

HIND WINGS: The two back wings of a four-winged insect

HOST: A plant or animal that a parasite feeds on. Typically, the host is somewhat injured or disabled. In extreme cases the host can die

INDICATOR SPECIES: An organism whose presence or absence is used to measure the quality of an environment

LARVA: The first stage of an insect's life after it leaves the egg during complete metamorphosis

MANDIBLES: A pair of mouthpart appendages, or jaws, used mainly for tearing and chewing food and carrying objects in insects and some other arthropods

METAMORPHOSIS: A change in structure from one body form to another, during a creature's life

MOULT: The process in which an insect sheds or loses a covering of skin, hair, feathers, etc.

NECTAR: A sweet liquid secreted by plants as food to attract animals that will benefit from them

NYMPH: The first stage of an insect's life after it leaves the egg during incomplete metamorphosis

OVIPOSITOR: The body part that female insects use to lay eggs

PARASITE: An organism that lives on or inside another species of organism (the host) and feeds on it

POLLEN: Tiny grains produced by the male part of flowers that fertilise the future seeds of a plant of the same species

PREDATOR: An animal that hunts other animals for food

PROBOSCIS: The elongated tubular mouthparts that some insects use to drink a liquid meal

PRONOTUM: A hardened plate on the top of the thorax just behind the insect's head that is part of the insect's exoskeleton

PUPA: A life stage of insects with complete metamorphosis during which the larval body is replaced with an adult body. In some insects, the pupa is enclosed in a cocoon

THORAX: The part of the body between the head and the abdomen. In insects, the wings and legs are attached to the thorax

Index

1st Edition
Libby Romero, *Author*

The publisher would like to thank the following members of the project team: Kevin Mulroy, Barbara Brownell Grogan, William Lamp, Matt Propert, Jane Sunderland, and Tim Griffin.

Art Directed by Kathryn Robbins

Designed by Chris Mazzatenta

British English edition
Michelle I'Anson, *Publishing Manager*
Keith Moore, *Editorial Lead*
Mark Steward, *Design and layout*
Shelley Teasdale, Karen Midgley, Beth Franklin, *Editors*

Original text, imagery and content oversight:
Paul Hetherington, Bug*life*

Published by Collins

An imprint of HarperCollins Publishers
Westerhill Road, Bishopbriggs, Glasgow G64 2QT
www.harpercollins.co.uk

HarperCollins Publishers
Macken House, 39/40 Mayor Street Upper, Dublin 1
D01 C9W8 Ireland

In association with National Geographic Partners, LLC

NATIONAL GEOGRAPHIC and the Yellow Border Design are trade-marks of the National Geographic Society, used under license.

Second edition 2020, First published 2017

ISBN 978-0-00-837455-6
10 9 8 7 6 5 4

A catalogue record for this book is available from the British Library

Printed in India.

If you would like to comment on any aspect of this book, please contact us at the above address or online.
natgeokidsbooks.co.uk
collins.reference@harpercollins.co.uk

Since 1888, the National Geographic Society has funded more than 12,000 research, exploration, and preservation projects around the world. The Society receives funds from National Geographic Partners, LLC, funded in part by your purchase. A portion of the proceeds from this book supports this vital work. To learn more, visit natgeo.com/info.

Photo Credits
The following images courtesy of **Bug*life*.** Contributing photographers named below:

page 4 (top) © Suzanne Burgess, 4 (mid) © Alan Stubbs, 4 (bot) © Paul Brock, 5 (top) © Steven Falk, 5 (bot) © David Pryce, 7 (top) © Steven Falk, 7 (top) © DenisG, 8 (bot) © Denis G, 9 (egg) © Buglife, 10 (top) © Steven Falk, 10 (bot) © Steven Falk, 12 © Steven Falk, 13 © John Walters, 15 (top) © Steven Falk, 15 (bot) © Roger Key, 16 (top) © Steven Falk, 17 (top) © Stephen and Alex Hussey, 17 (bot) © Jon Mold, 18 (top) © Francis Rowland, 18 (bot) © Ben Hamers, 19 (bot) © Jay Steel, 20 (top) © Suzanne Burgess, 20 (bot) © John Walters, 21 (bot) © Roger Labbett, 22 (bot) © Buglife, 23 (top) © Steven Falk, 23 (bot) © Ben Hamers, 24 (top) © Steven Falk, 24 (bot) © Jon Mold, 25 (top) © Steven Falk, 25 (bot) © Dan TP, 26 (top) © Suzanne Burgess, 26 (bot) © Steven Falk, 27 (top) © Dr David Chesmore, 27 (bot) © Roger Key, 28 (top) © Ben Hamers, 28 (bot) © Suzanne Burgess, 29 (top) © Matt Shardlow, 29 (bot) © Andrew Whitehouse, 30 (top) © Chris Gibson, 30 (bot) © Iain Perkins, 31 (top) © Greg Hitchcock - www.grhphotography.co.uk, 31 (bot) © Bill Urwin, 32 (top) © Steven Falk, 32 (bot) © Ben Hamers, 33 (top) © Roger Key, 33 (bot) © Gelena Wilby, 34 (top) © Steven Falk, 35 (top) © John Walters, 37 (top) © Steven Falk, 38 © Steven Falk, 39 (top) © Alan Stubbs, 40 (top) © Rob Mills, 40 (bot) © Steven Falk, 41 (top) © Suzanne Burgess, 41 (bot) © Liza Fowler, 42 (top) © Liza Fowler, 42 (bot) © Steven Falk, 43 (top) © Steven Falk, 43 (bot) © Alan Stubbs, 44 (top) 44 © Rob Garrod, 45 (top) © Ian Dawson, 45 (bot) © Rory Dimond, 46 (top) © Steven Falk, 46 (bot) © Dan TP, 47 (top) © Claudia Watts, 50 (top) © Raz www.phocus-on.co.uk, 50 (bot) © Stuart Crofts, 51 (top) 51 © Jon Mold, 51 (bot) © Roger Key, 52 (bot) © Roger Key, 53 (bot) © RMS, 54 (bot) © Jaroslav Maly, 55 (top) © Paul Brock, 56 (top) © Roger Key, 56 (bot) © Roger Key, 57 © Steven Falk, 58 (top) © Buglife, 58 (bot) © Dave Eagle, 59 (top) © Suzanne Burgess, 60 (top) © Roger Key, 61 (top) © Roger Key, 61 (bot) © Buglife, 64 (bot) © Dr Bernhard Seifert, 65 (top) © Gus Jones, 66 (top) © Steven Falk, 66 (bot) © Stuart Crofts, 67 (top) © Phil Clayton, 67 (bot) © Suzanne Burgess, 68 (top) © Steven Falk, 68 (bot) © Steven Falk, 70 (top) © Steven Falk, 70 (bot) © G Hiscocks, 71 (top) © Steven Falk, 72 (top) © Steven Falk, 72 (bot) © Steven Falk, 73 (top) © Suzanne Burgess, 74 (top) © John Mason, 75 © Nick Packham, 76 (top) © Steven Falk, 76 (bot) © Steven Falk, 77 © Steven Falk, 79 (top) © Steven Falk, 79 (bot) FalkBuglife, 80 (top) © Steven Falk, 80 (bot) © Richard Donoyou, 81 © Roger Key, 84 (top) © Steven Falk, 84 (bot) © Steven Falk, 85 (top) © Iain Perkins, 85 (bot) © Steven Falk, 86 (top) © Steven Falk, 86 (bot) © Steven Falk, 87 (top) © Iain Perkins, 87 (bot) © Steven Falk, 88 (top) © Steven Falk, 89 (top) © Steven Falk, 90 (top) © Steven Falk, 91 (top) © Scott Shanks, 92 (top) © Steven Falk, 92 (bot) © Steven Falk, 93 (top) © Paul Hetherington, 94 (top) © Steven Falk, 95 (top) © Dan TP, 95 (bot) © Roger Key, 96 (top) © Dan TP, 96 (bot) © Dan TP, 97 (top) © Darren Bradley, 98 (top) © Iain Perkins, 98 (mid) © Simon Munnery, 99 (top) © Stig Madsen, 102 (top) © Steven Falk, 102 (bot) © Roger Key, 103 © Wildfeuer, 104 (top) © Suzanne Burgess, 104 (mid) © Roger Key, 104 (bot) © Buglife, 105 (top) © Steven Falk, 105 (bot) © Steven Falk, 106 (top) © Steven Falk, 106 (bot) © Gareth rondel, 107 (bot) © Steven Falk, 108 (top) © Steven Falk, 109 (top) © Steven Falk, 109 (bot) © David Pryce, 110 (top) © Paul Hetherington, 110 (bot) © Steven Falk, 111 (top) © Roger Key, 112 (top) © Steven Falk, 112 (bot)© Steven Falk, 113 (top) © Steven Falk, 113 (bot) © Ursula Smith, 114 (top) © Malcolm Lee, 114 (bot) © Malcolm Lee, 115 © Craig Macadam, 116 (top) © Suzanne Burgess, 117 (top) © Alan Stubbs, 118 (top) © Stuart Crofts, 121 © Roger Key

Other images courtesy of SHUTTERSTOCK, or in public domain

MIX
Paper | Supporting responsible forestry
FSC
www.fsc.org
FSC™ C007454